Ramy Bishay

Seeing the Transplanted Mouse Heart Like Never Before

Ramy Bishay

Seeing the Transplanted Mouse Heart Like Never Before

New Insights Into Experimental Heart Rejection
Using Novel Ultrasound Applications

VDM Verlag Dr. Müller

Imprint

Bibliographic information by the German National Library: The German National Library lists this publication at the German National Bibliography; detailed bibliographic information is available on the Internet at http://dnb.d-nb.de.

Any brand names and product names mentioned in this book are subject to trademark, brand or patent protection and are trademarks or registered trademarks of their respective holders. The use of brand names, product names, common names, trade names, product descriptions etc. even without a particular marking in this works is in no way to be construed to mean that such names may be regarded as unrestricted in respect of trademark and brand protection legislation and could thus be used by anyone.

Cover image: www.purestockx.com

Publisher:
VDM Verlag Dr. Müller Aktiengesellschaft & Co. KG, Dudweiler Landstr. 125 a, 66123 Saarbrücken, Germany,
Phone +49 681 9100-698, Fax +49 681 9100-988,
Email: info@vdm-verlag.de

Produced in USA and UK by:
Lightning Source Inc., La Vergne, Tennessee, USA
Lightning Source UK Ltd., Milton Keynes, UK
BookSurge LLC, 5341 Dorchester Road, Suite 16, North Charleston, SC 29418, USA

ISBN: 978-3-8364-8738-2

ABSTRACT

HIGH-FREQUENCY ULTRASOUND ASSESSMENT OF CARDIAC GRAFT REJECTION IN HETEROTOPICALLY TRANSPLANTED MICE

Rami Bishay

Master of Science 2007

Department of Physiology, Faculty of Medicine, University of Toronto

We explored the utility of ultrasound biomicroscopy (UBM) as an improved approach over the common method of finger palpation for assessing graft rejection in murine heterotopic cardiac grafts *in vivo*. Syngeneic (C3H->C3H, n=9) and allogeneic (BALB/c->C3H, n=13) grafts were imaged daily using a 30 MHz transducer for 14 days post-transplant or until rejection (median survival time=12 days), respectively. Mean palpation score for allografts significantly differed from isografts on day 7 post-transplant, whereas UBM detected significantly increased left ventricular (LV) wall thicknesses from day 2 onwards and LV dilation from day 10 post-transplant in allografts ($P \leq 0.001$), but not isografts, which is likely due to myocardial edema and immune-mediated rejection. Novel Doppler waveforms of aortic and coronary flow revealed changes associated with LV dilation and diminishing systolic function in rejecting allografts. UBM-derived measurements of LV wall thicknesses provide a non-invasive method for detecting differences between isografts and allografts earlier than finger palpation.

ACKNOWLEDGEMENTS

There were many personal and professional challenges during my stint as a master's student, from which this monograph was bourne, and it is clear that I could not have overcome these obstacles if it were not for the support of my supervisors, colleagues, and family. I'd like to very much thank my supervisor Dr. Lori West for recruiting me to her successful laboratory as well as for her continual support and concern for the professional development of her students. Dr. Lee Adamson, my co-supervisor, was instrumental in the continuation of this project and my degree during the West lab's migration outside the province. Her attention to detail, constructive criticisms, and amiable demeanour have made the lab's transition personally seamless. This work was also made possible by the enduring efforts of Dr. Yu-qing Zhou from the Mouse Imaging Centre (MICe), whose mastery of ultrasound imaging and establishment of an imaging method in grafted mice was critical to my own success. Thank you for being patient throughout my 'learning curve.' I am grateful to Dr. Mark Henkelman, head of the MICe, who allowed me to acquire professional training in 'cutting edge' imaging and for opening the MICe to me. I'd also like to recognize the efforts of my graduate committee, Dr. Thomas Waddell and Dr. Mansoor Husain, exceptionally busy clinician-scientists who made the time to lend their critical appraisal for the benefit of this project.

Lydia Mai and Kesheng Tao, past and present members of the West lab, respectively, were excellent exemplars of reliability and support. I have yet to recall a time when they were not willing to undertake long hours of microsurgery for this project. Other notable members of the West lab include Andrew Ang, Dr. Mylvaganam Jeyakanthan, Dr. Frank Fan, and summer students and future physicians, Katie Marchington and Yang Li, who made the summer of 2005 one to remember. I would like to show gratitude to Wei Zhu for his generosity with his time in facilitating procurement of the ECG data. Shathiyah Kulandavelu and John Sun of the Adamson lab were also exceptionally helpful in relating their experiences in biological imaging techniques and were valuable in welcoming me into the Adamson lab. I'd like to also acknowledge the help of Lori Davidson and Shoshana Spring, members of the MICe, who made those long afternoons of imaging unusually pleasant and helped me locate reagents in a place that quickly became my 'second home' for six months during the data acquisition phase. The congenial graduate students of the Adeli lab who formed the grad lounge gang must be mentioned also; I wish them the best in their academic and personal aspirations.

My parents were inarguably the most involved in every step of my life and they continued that same precedent during the 'ups-and-downs' in the last several years. Their love and support will not be forgotten for as long as I live. My brother Ray has always been the voice of reason; he helped me to see the proverbial 'light at the end of the tunnel' when one was hard to find. My younger sister Christine's gift for sporadic yet often hilarious behaviour was particularly valuable to me, as she scarcely left my presence without making me laugh. My long-time friends, Peter, Mena C, and Remon were always available for moral support and I'm grateful for their willingness to listen. Since my move to Sydney, Australia for yet another degree, I've made many close friends who quickly have become my surrogate family many miles away from Toronto; they include Jason, Karim, John M, Mena S, and John R.

The blessings, love, and strength of the Lord were, are, and will continue to be the crux that holds all things in harmony in my life. I am where I am because of God and for that, I am eternally indebted and grateful.

TABLE OF CONTENTS

LIST OF ABBREVIATIONS

2D	Two-dimensional
AA	Ascending aorta
ACR	Acute cellular rejection
AI	Aortic insufficiency
ALLO	Allograft
Apo	Apolipoprotein
AR	Aortic regurgitation
AWes	Anterior wall at end-systole
AWed	Anterior wall at end-diastole
BPM	Beats per minute
Bwd	Backward
B-mode	Brightness mode
CO	Cardiac output
ECG	Electrocardiography
EDD	End-diastolic diameter
EF	Ejection fraction
EKV	ECG based kilo hertz visualization
ESD	End-systolic diameter
FS	Fractional shortening
Fwd	Forward
H&E	Haematoxylin and eosin
HHTx	Heterotopic heart transplant
HLA	Human leukocyte antigen
HR	Heart rate
ICAM-1	Intercellular adhesion molecule-1
IFN	Interferon
IL	Interleukin
I/R	Ischemia-reperfusion
ISO	Isograft

IVC	Inferior vena cava
IVRT	Isovolumic relaxation time
LA	Left atrium
LAV	Long-axis view
LCA	Left coronary artery
LDL	Low-density lipoprotein
LV	Left ventricle
mAb	Monoclonal antibody
MHC	Major Histocompatibility complex
MHz	Mega hertz
MR	Mitral regurgitation
M-mode	Motion mode
MPA	Main pulmonary artery
MRI	Magnetic resonance imaging
mRNA	Messenger ribonucleic acid
MPTP	Mitochondrial permeability transition pore
NMR	Nuclear magnetic resonance
PTx	Post-transplant
PWed	Posterior wall at end-diastole
PWes	Posterior wall at end-systole
RCA	Right coronary artery
RI	Resistance index
RV	Right ventricle
SAV	Short-axis view
SV	Stroke volume
T_E	Relaxation time constant
TNF	Tumour necrosis factor
TUNEL	Terminal deoxynucleotidyl transferase biotin-dUTP nick end labelling
TVI	Time-velocity integral
UBM	Ultrasound biomicroscopy

Ved	End-diastolic velocity
Vmax	Maximum velocity

LIST OF FIGURES

LIST OF TABLES

I. INTRODUCTION

1.0   INTRODUCTION SUMMARY AND RATIONALE

1.0.1   Clinical Problem

Medical science owes much of its progress to the use of animal models of human disease. In particular, given the rapid growth in genetic engineering, mouse models have become a common and ideal tool to study the aetiology and pathogenesis of several diseases, such as cancer (Nomura *et al.*, 2006), diabetes (Planas *et al.*, 2006), genetic disorders (Juncos *et al.*, 2006) and graft rejection (Kosuge *et al.*, 2006). Cardiac graft rejection, for example, is a major problem in clinical transplantation. Despite advances in the development of specific and potent immunosuppressants (Morris, 1993), our understanding of the immunopathology of transplantation, tissue typing, and organ preservation, management of acute rejection remains poor (Petri, Jr., 1994). Acute cellular rejection (ACR) is the leading cause of mortality in heart transplant recipients, amounting to 20% of deaths in the first year post-transplant (PTx) (Hosenpud *et al.*, 2001). The picture is much bleaker when one considers that by 5 years PTx, diagnosis of cardiac allograft vasculopathy – a form of accelerated atherosclerosis seen uniquely in transplant patients – reaches 50% of transplant recipients (Billingham, 1992). Thus, investigations into elucidating mechanisms of rejection and in ameliorating graft rejection are on-going.

1.0.2   The Heterotopic Heart Transplant (HHTx) Mouse Model

The murine heterotopic heart transplantation (HHTx) model was first introduced about three decades ago for the purposes of studying the roles of major histocompatibility complexes in cardiac graft rejection (Corry *et al.*, 1973b). Today, our laboratory as well as others have established proficiency in heterotopic cardiac transplantation in mice, due to advances in microsurgery techniques that have allowed development of appropriate models of primarily-vascularized (or fully vascularized) organ transplantation in small rodents (Niimi, 2001). Cardiac allografts and isografts are harvested at various times after transplantation for histologic examination for evidence of rejection. Therapeutic interventions that may prevent or delay rejection can then be studied. This is particularly useful since many genetically well-characterized inbred mouse strains are available. The possibilities are even more appealing when one considers the plethora of novel transgenic

1

and gene knock-out mice strains that are continually being generated; specific genetic disparities therefore have the potential to become powerful tools for elucidating the molecular processes underlying graft rejection (Krieger and Fathman, 1997), (Yamada *et al.,* 2003), for exploring molecular approaches of improving inducing graft tolerance and acceptance (West and Tao, 2002; West *et al.,* 1994) and for testing of pharmacological agents on graft rejection and acceptance (Kosuge *et al.,* 2006; Mottram *et al.,* 1993).

1.0.3   Assessing Graft Rejection in the Murine HHTx Model

Despite the powerful utility of this model to our understanding of immune mechanisms of cardiac graft rejection, no qualitative or quantitative descriptions of graft function have been established in this model. The current method for determining the status of the rodent HHTx graft involves daily finger palpations of the graft to determine the force of contraction; a scale between '4' (vigorous palpable cardiac activity) and '0' (cessation of palpable cardiac activity) is then assigned by the observer (Corry *et al.,* 1973b). This crude method of assessment introduces a high level of inter-observer variability, is insensitive to various degrees of rejection, and yields exceptionally limited information on long-term graft survival and function (Mottram *et al.,* 1988). Biopsy remains the 'gold standard' for assessing rejection in clinical surveillance though it is invasive and inappropriate for mice (Caves *et al.,* 1973; Billingham *et al.,* 1973). Needle electrodes can yield electrocardiogram traces of heterotopic grafts, but are affected by location of the electrodes, the variable orientation of grafts, (Mottram *et al.,* 1988) and electrical interference from surrounding structures such as the intestines (Zhou *et al.,* 2007). Other methods that employ molecular biomarkers (Tanaka *et al.,* 2005d), NMR spectroscopy (Haug *et al.,* 1987), MRI (Kanno *et al.,* 2001) and scintigraphic imaging (Ohtani *et al.,* 1995) have been limited by high cost, requirement for radiological materials, and lack of information that can be gained pertaining to the physiology of the graft during the rejection process.

1.0.4   Ultrasound Biomicroscopy (UBM)

Developed for clinical ophthalmic and dermatological applications (Pavlin *et al.,* 1990; Sherar *et al.,* 1989), ultrasound biomicroscopy (UBM) - often referred to as high-frequency ultrasound - has developed into a powerful tool to study cardiac function in mice

2

(Zhou *et al.*, 2004; Foster *et al.*, 2000). Previously, clinical echocardiography systems have been employed and have resulted in observations of increased ventricular wall thicknesses in murine heterotopic cardiac allografts (Scherrer-Crosbie *et al.*, 2002), though limited spatial resolution at lower clinical frequencies (8-12 MHz) may not be appropriate for detailed evaluations of graft physiology and haemodynamic patterns in the small mouse heart (Zhou *et al.*, 2007). Recently, *in vivo* cardiac imaging in mice using high frequencies (20-55 MHz) has been used to evaluate cardiac dimensions, function and haemodynamics with much higher spatial resolution (Zhou *et al.*, 2004). Therefore, it became apparent that this imaging modality has the potential to offer significant advantages over previous methods of palpation, histology and other methods (discussed below) and is capable of yielding valuable information previously undocumented in this model of cardiac transplantation. To that end, we proposed to evaluate the physiologic, haemodynamic and morphological changes in syngeneic and allogeneic transplants over the post-transplant course serially using the novel UBM system. Graft function was assessed using a battery of echocardiographic parameters established for human echocardiography that were modified by our laboratory for this model system. The fruits of this work would help delineate the particular changes allografts and isografts undergo in response to immunological rejection (or lack thereof) and potentially give rise to sensitive and specific UBM-derived predictors of rejection in this model. Furthermore, other information not fully reported in this common experimental model, such as the effects of ischemia-reperfusion injury, cardiac unloading, and the interaction of native and graft cardiac cycles, will put into context results obtained from past and future research using this murine model.

1.1 CLINICAL CARDIAC TRANSPLANTATION

1.1.1 Introduction

Heart transplantation is scarcely recognizable from what it was when Dr. Christiaan Barnard performed the world's first heart transplant on December 3, 1967 in Cape Town, South Africa (Baxter and Smerdon, 2000). Advances in surgical techniques, preservation protocols and pharmacological treatments in the past two decades have seen a considerable improvement in short-term patient survival following cardiac transplantation, with reported survival rates of 80%-90% after one year PTx (Bourge *et al.*, 1993). These favourable

3

outcomes have been largely attributed to the introduction of more potent and specific immunosuppressants, such as cyclosporine and tacrolimus (Morris, 1993), improvements in the understanding of the immunology of transplantation, tissue typing, organ preservation, and management of acute rejection and opportunistic infections (Petri, Jr., 1994). With more than 31,000 heart transplants performed to date, cardiac transplantation is an established and preferred choice of treatment for patients with congenital heart defects or severe end-stage congestive heart failure in whom maximal medical therapy has failed (Hosenpud *et al.,* 2001). Dilated cardiomyopathy and ischaemic heart disease are the most common indications for heart transplantation (~40% each) (Dengler and Pober, 2000).

1.1.2   Problems in Clinical Transplantation
1.1.2.1 Supply and Demand

Despite considerable advances, there remains a staggering worldwide gap between the supply and demand for transplantable organs, and the gap is enlarging. Almost half of all patients on the waiting list for a heart graft will never receive one (Cooper *et al.,* 2000). The problem in Canada has reached a level akin to a medical crisis. In 1999, there was a 21% increase in patients on the waiting list for a transplant, but only a 6% increase in eligible heart donors over the same period, with a consequent annual waiting mortality rate of 25% (CHI Registry Data, 2006). Generally, the number of organ donors (14-15 donors per million pop.) has remained fairly constant each year in Canada (CIHI Report, 2004).

There are currently several lines of investigation into alternative therapies for patients suffering from cardiomyopathy and end-stage heart failure. Xenotransplantation, for example, could potentially provide an unlimited number of non-human donors, although this approach is limited by fulminant graft rejection mediated by pre-formed antibodies and complement activation (Cooper *et al.,* 2000). Research into stem cell replacement therapy is aimed at regenerating the tissue of damaged portions of myocardium due to congestive heart failure or following myocardial infarction and have been purported to ameliorate heart function in mice and rats (Shepler and Patel, 2007). Similarly, engineering of heart tissues *in vitro* for implantation onto the failing heart have been shown to improve systolic function in animal models, but constraints on the size of the generated tissue as well as the lack of sustained contractile properties of the engineered tissue remain obstacles (Zimmermann *et*

4

al., 2006a; Zimmermann *et al.*, 2006b). Gene therapy has also been explored in rodent models as a method for the viral transfer of gene products into the donor orgran during preservation that possess anti-inflammatory and suppression of T cell function capabilities (Isobe *et al.*, 2004), however, this treatment strategy is far from being validated for clinical use. Thus, at present, cardiac transplantation remains the only definitive treatment option for patients with end-stage heart failure.

1.1.2.2   *Graft Rejection*

The success of cardiac transplantation as an established medical therapy is also substantially restricted by graft rejection, which is due to the potentially devastating response of the immune system to transplanted tissue. Of the four major patterns – hyperacute, acute cellular, acute vascular and chronic – acute cellular rejection (ACR) remains the leading cause of mortality in the first year post-transplantation in heart transplant recipients, amounting to 20% of deaths (Hosenpud *et al.*, 2001). ACR can occur as early as 2-3 weeks and as late as several years after transplant, though it is most common in the first post-operative year (Dengler and Pober, 2000). Despite maintenance immunosuppression, 50-70% of cardiac transplant patients will have at least one significant episode of ACR that requires treatment (Hosenpud *et al.*, 2001).

Long-term graft viability is commonly hindered by chronic rejection, which is manifested typically in the form of cardiac allograft vasculopathy; other pseudonyms for cardiac allograft vasculopathy include graft arteriosclerosis, accelerated arteriosclerosis, and allograft vasculopathy disease (Weis, 2002). Though the incidence of cardiac allograft vasculopathy is 5% to 10% of new cases per year of the post-operative period, intimal thickening, a classic early sign of the pathogenesis of the disease, is present in 58% of transplant patients during the same year after transplantation (Billingham, 1992). By 5 years, diagnosis of cardiac allograft vasculopathy reaches 50% in adult transplant recipients (Moien-Afshari *et al.*, 2003).

The much more aggressive humoral response of hyperacute rejection is caused by pre-formed antibodies to blood group (ABO) and human leukocyte antigens (HLA). It occurs within minutes to hours after transplant and results in immediate malfunctioning of the graft. Hyperacute rejection is characterised by widespread thrombotic occlusion of graft

5

vasculature mediated by pre-existing antigraft antibodies and activation of the complement cascade (Dengler and Pober, 2000).

1.1.2.3   Acute Cellular Rejection (ACR)

ACR remains the leading cause of mortality in heart transplant recipients 1 year post-transplantation. Histologically, ACR is characterised by infiltrating host lymphocytes and macrophages, commonly found in either the perivascular cuffs or in the myocardium. Necrotic myocytes are also common and found typically associated near cytotoxic T lymphocytes. Cytotoxic T cells are the potent killing agent of the immune system, attacking foreign, diseased, and infected cells. They are appropriately referred to as killer T cells, carry the CD8 marker and recognize major histocompatibility complex (MHC) Class I molecules (Dengler and Pober, 2000).   Though neutrophil infiltration (Healy *et al.*, 2006), CD4+ cells, and natural killer cells (Pietra and Gill, 2001) are involved in the immunological response against foreign graft tissue (Grazia *et al.*, 2004), studies suggest that host CD8+ cells are sufficient alone to mediate allograft rejection (Kreisel *et al.*, 2002). Host cytotoxic T cells recognize alloantigens presented by donor antigen-presenting cells bearing MHC Class I receptor – which are found ubiquitously in all nucleated somatic cells – normally via direct presentation (*i.e*, antigen presenting cells from the donor presents foreign/donor antigen to host cytotoxic T cells) in the donor graft vascular endothelium. Subsequent to binding of donor MHC Class I molecule with the T cell receptor – an antigen specific membrane receptor – cytotoxic T cells generate a molecular cascade resulting in the expression of perforin and granzyme B from the T cell resulting in apoptosis and consequent lysis of the target donor graft cell (Alpert *et al.*, 1995). Activated CD8+ cells differentiate into effector (killing) and memory cells with the latter secreting interferon-☐

In contrast to CD8+ cells, host CD4+ cells recognize MHC Class II molecules and can undergo direct (via donor antigen presenting cells) or indirect allorecognition (*i.e.,* via host antigen presenting cells, typically macrophages and dentritic cells). CD4+ allorecognition results in CD4+ differentiation into helper T cells, clonal expansion, and release of cytokines (*e.g.,* interleukin-2, interferon-☐). Cytokines up-regulate clonal expansion of CD8+ cells, B cells, and macrophages, the latter being critical for generating an inflammatory response (Suzuki *et al.*, 1998; Pietra and Gill, 2001). Both endothelial and

6

myocardial cells can be destroyed through the lytic action of activated macrophages, resulting in endothelialitis, and haemorrhage (Dengler and Pober, 2000). Subsequently, damaged or dysfunctional endothelium secretes adhesion molecules, integrins, and chemokines, which facilitate binding and transendothelial migration of T cells, where they can interact with antigen presenting cells and cause tissue destruction (Blake, 2004).

1.1.2.4   Hyperacute Rejection

In contrast to acute vascular rejection, which involves a cellular and antibody component, the role of antibody-secreting B lymphocytes and complement activation in the immune response to graft tissue is much more prominent in hyperacute rejection. A variety of clinical and histopathological patterns are seen and they all generally lack a cellular component (Dengler and Pober, 2000). Subsequent to binding of a B cell receptor (a membrane-bound antibody) with foreign intact proteins, nucleic acids, polysaccharides, or glycolipids, the antigen receptor initiates a sequence of events that leads to B cell maturation into antibody-producing plasma cells. Complement proteins bind to alloantigen-bound antibodies and catalyze the formation of a membrane attack complex, resulting in apoptosis and destruction of target cells. The process also recruits inflammatory cells. Ischaemic endothelium during the perioperative period may also stimulate complement activation (Dengler and Pober, 2000).

Production of antibodies to protein antigens is facilitated by cytokine secretion and T helper cells (therefore termed T-dependent) whereas antibodies formed against polysaccharides for example are made in a T independent manner. The latter immunologic response is a common limitation for donor organ availability of ABO-mismatched recipients, though intense research has validated a clinical protocol for ABO-mismatched transplantation in neonates due to immunologic immaturity (Fan *et al.*, 2004; West *et al.*, 2001).

1.1.2.5   Chronic Rejection

Chronic rejection commonly results in fibrosis, or as a progressive immune reaction due to low immunogenicity (Dengler and Pober, 2000). As previously mentioned, it is commonly manifested as cardiac allograft vasculopathy. Though the pathogenesis of

7

allograft vasculopathy is poorly understood, the consensus is that the disease is caused by immunologic mechanisms (*e.g.,* alloreactive T cells and the humoral immune system), non-immunologic mechanisms relating to the transplant itself or the recipient (*e.g.,* hypertension, hyperlipidemia and pre-existing diabetes) or to the side effects often associated with immunosuppression by calcineurin inhibitors or corticosteroids (*e.g.,* cytomegalovirus infection, nephrotoxicity and new-onset diabetes) (Weis, 2002). Only donor graft vessels are susceptible to the development of cardiac allograft vasculopathy. Studies however point to the critical role of the endothelium in the development of allograft vasculopathy, which is capable of releasing pro-fibrotic cytokines and recruiting leukocytes. The resultant luminal narrowing of graft vessels through the formation of an expanding neointima by proliferation of smooth muscle cells in the vessel wall and deposition of matrix proteins ultimately leads to declining graft function (Weis, 2002; Dengler and Pober, 2000).

1.1.3   Clinical Echocardiographic Features of Cardiac Grafts
1.1.3.1   Tricuspid Regurgitation
Tricuspid regurgitation is the most prevalent valvular dysfunction in orthotopic grafts and in one study, occurs in about 20% of transplant patients after final separation from cardiopulmonary bypass (Anderson *et al.,* 2004). It is primarily due to elevated pulmonary vascular resistance and secondary ventricular dysfunction. Tricuspid regurgitation has also been shown to be associated with poor long-term graft survival (Anderson *et al.,* 2004).

1.1.3.2   Left Ventricular Dysfunction and Myocardium
Left ventricular (LV) diastolic filling is influenced by loading conditions (*e.g.,* preload and afterload changes), ventricular hypertrophy, acute rejection, and myocardial fibrosis (Peteiro *et al.,* 1996). Early LV systolic dysfunction is usually due to injuries incurred during preservation, and an increase in LV mass and mural thickness shortly after transplantation represent the effects of myocardial ischemia and edema and are independent of rejection. Later acute increases in left ventricle mass and myocardial thickness, however, may be attributable to rejection (Sagar *et al.,* 1981).

1.1.3.3   The Right Ventricle

In the first few days after transplantation, right ventricular dilation may occur due to elevated pulmonary vascular resistance, though a progressive decline is seen over the subsequent weeks. Right ventricular ejection fraction remains preserved in the majority of cases (Bhatia *et al.*, 1987; Greenberg *et al.*, 1985).

1.1.3.4   Pericardial Effusion

Pericardial effusions can be observed early post-transplant and are primarily due to the differences in size between the large failing explanted heart and the implanted heart (Weitzman *et al.*, 1984). Echocardiography can serially follow the progress of pericardial fluid. Post-operative pericardial effusions typically resolve with time and are not usually a manifestation of acute rejection (Vandenberg *et al.*, 1988), though late effusion often represents a noninfectious or inflammatory reaction within the epi- or pericardium and may reflect ACR.

1.1.3.5   Cardiac Size and Heart Rate (HR)

In the early years of cardiac transplantation (early 70s), an increase in cardiac size as determined by echocardiography was correlated with rejection (Griepp *et al.*, 1971), but this observation was too non-specific to be of any routine clinical value. Instead, HR variability has been studied and found to be associated with severe rejection, with a sensitivity and specificity of 77% and 76%, respectively, though this method is used only as an adjunct to biopsy (Sands *et al.*, 1989; Zbilut *et al.*, 1988).

1.1.3.6   Myocardial Mass

Another prominent clinical feature of rejecting cardiac grafts is an increase in myocardial mass due to inflammatory cell infiltration and myocardial edema, which is clearly recognizable by 2-dimensional (2D) imaging (Burgess *et al.*, 2002). Unfortunately, measurable changes in the myocardium based on M-mode or 2D imaging methods occurred later than a biopsy-proven episode of rejection. Furthermore, since the advent of cyclosporine therapy, rejection episodes are associated with less myocardial edema in

9

patients during episodes of ACR than non-cyclosporine treated patients of the 60's and 70's, and evaluating myocardial mass is considered obsolete (Burgess *et al.*, 2002).

1.1.4 Echocardiography and Cardiac Rejection

1.1.4.1 Problems with Endomyocardial Biopsy

Endomyocardial biopsy has remained the procedure of choice for surveillance monitoring of transplant patients (Burgess *et al.*, 2002). In 1990, an international grading system for cardiac allograft biopsies was adopted by the International Society for Heart Transplantation but was revised in 2004 to address challenges and inconsistencies in its use and to address recent advances in the knowledge of antibody-mediated rejection (Stewart *et al.*, 2005). The revised (R) categories of cellular rejection are currently as follows: Grade 0 R: no rejection (no sign of rejection); Grade 1 R: mild rejection (interstitial and/or perivascular lymphocytic infiltrate with up to 1 focus of myocyte damage); Grade 2 R: moderate rejection (two or more foci of infiltrate with associated myocyte damage); and Grade 3 R: severe rejection (diffuse infiltrate with multifocal myocyte damage ± edema, ± haemorrhage ± vasculitis) (Stewart *et al.*, 2005). Despite these advancements in diagnosis, however, the distribution of rejection within the myocardium may be non-uniform and even with multiple biopsies from different sites, the grade of rejection in the myocardium as a whole may be underestimated (Nakhleh *et al.*, 1992). For noninvasive imaging to replace biopsy, the sensitivity for the detection of rejection would need to approach 100% (Burgess *et al.*, 2002).

1.1.4.2 Systolic and Diastolic Dysfunction

Decreases in left ventricular %fractional shortening (%FS), stroke volume (SV), and %ejection fraction (%EF) have been reported to correlate with the degree of rejection, though they are generally deemed not specific echocardiographic predictors of graft rejection since similar but less remarkable changes in %FS, SV and %EF can also occur during other events, such as systemic infection and immediate post-operative haemodynamic compromise despite no rejection (Sade *et al.*, 2006). Because abnormalities of diastolic filling generally occur before any evidence of systolic dysfunction (Paulsen *et al.*, 1985), evaluating graft diastolic dysfunction using echocardiography has the potential for detecting rejection non-invasively.

Currently, Doppler parameters of diastolic function in most transplant centers are an adjunct to rather than replacement for biopsy. Studies show that preserved Doppler parameters of diastolic function early post-transplant are associated with a higher late-term actuarial survival (Ross et al., 1996). Some parameters that have been proposed include: decreased mitral deceleration time of the early filling phase or shortening of isovolumic relaxation time (IVRT) of at least 15% (Valantine et al., 1987); use of acoustic quantification echocardiography to monitor a an 18% decrease in peak filling rate (Moidl et al., 1999); endocardial border detection of increased left ventricular filling during diastasis, and decreased filling during the rapid filling phase (Kimball et al., 1997); decreased mitral-A wave velocity transit time from left ventricular inflow to outflow tracts (Boyd et al., 1997); and Doppler tissue imaging to assess reduced myocardial relaxation velocities (Puleo et al., 1998). Alternatively, myocardial performance index, also known as the Tei index, is a relatively new method of measuring global systolic and diastolic function. It is equal to the sum of the isovolumic times divided by the ventricular ejection time. The index is independent of heart rate, load, and requires no geometric assumptions, making it suitable for measuring ventricular function of cardiac grafts (Tei, 1995).

1.1.4.3   Cardiac Allograft Vasculopathy

Studies have established intravascular ultrasound as a more sensitive test than the previous method of angiography for the detection of early cardiac allograft vasculopathy (St Goar et al., 1992). Intravascular ultrasound provides characterization of vessel wall morphology and has been used to assess the degree of luminal stenosis common in the pathology of allograft vasculopathy as well as the amount of atheromatous plaque built up at any particular point in the epicardial coronary artery, which can be an accelerated process in patients suffering from the disease (St Goar et al., 1992). Intracoronary Doppler measurements of coronary flow velocity have also been used to evaluate graft vasculopathy and in heart transplant patients with preserved left ventricular function, a significantly abnormal coronary flow velocity reserve in the left anterior descending coronary artery indicated the severity of coronary artery disease and was associated with a high risk of cardiac events during an average of a 25 month follow-up post-transplant (Rodrigues et al., 2005). The predictive power of intracoronary Doppler assessments, however, has been

11

questioned in earlier studies (Klauss *et al.*, 1997). Some investigators have looked to evaluating endothelial dysfunction associated with the pathogenesis of cardiac allograft vasculopathy, as assessed by increases in mean annual change in area response and mean annual decrement in flow response - both evaluated following intracoronary infusion of acetylcholine; these measurements were found to be associated with patients who had angiographic evidence of allograft vasculopathy (Hollenberg *et al.*, 2004).

1.1.4.4   M-mode Analysis of Graft Rejection

As mentioned previously, increased ventricular mass and inflammatory edema that occur during graft rejection have been identified during episodes of ACR using 2D imaging though these observations have been rendered obsolete since cellular rejection in the era of cyclosporine therapy is associated with less edema and inflammation (Burgess *et al.*, 2002; Tei, 1995). Multi-parametric M-mode interrogation, however, which utilizes 2D guided-M-mode digital tracings of the LV, has shown that LV mass increased acutely from 109% of predicted normal values to 129% with rejection and decreased to 110% with therapy, whereas LV volume tended to fall with rejection and increase with therapy (Boucek *et al.*, 1993). Systolic function was depressed by rejection as reflected in the increased posterior wall thickening fraction and velocity of wall thickening and diastolic dysfunction was reflected in a decreased velocity of posterior wall thinning and depressed average velocity of cavity enlargement (Boucek *et al.*, 1993).

1.1.4.5   Stress Echocardiography

Dobutamine stress echocardiography has become a favoured technique to evaluate the presence of inducible myocardial ischemia and transplant coronary artery disease following heart transplantation (Pahl *et al.*, 1999; Larsen *et al.*, 1998). Evaluating the prognostic value of dobutamine stress echocardiography to predict acute cardiac events and mortality in heart transplant patients, Akosah *et al* showed that cardiac graft recipients followed for a mean of 2 years had a decreased %EF and a higher peak wall motion score index during cardiac events than in those without events, illustrating the potential of dobutamine stress echocardiography as a minimally invasive surveillance procedure for cardiac transplant patients (Akosah *et al.*, 1996).

A summary of parameters used to diagnose, evaluate, and/or associated with clinical acute cardiac rejection are outlined in **Table I**.

Table I. Summary of Clinical Cardiac Transplant using 2-D Echocardiography

2-D Clinical Echocardiographic parameters associated with acute cardiac graft rejection	Directionality and mechanism of change in parameter in clinical acute graft rejection
Graft mass J Am Soc Echocardiogr 2002;15:917-25.	Increased myocardial allograft mass due to inflammatory cell infiltration and myocardial edema in grafts undergoing rejection
Left ventricular mass Circulation. 1981 Aug;64(2 Pt 2):II217-20	LV mass increased 3x during episodes of rejection due to interstitial edema and cellular infiltrate
Aortic regurgitation Am J Cardiol. 1994;73:1197-201.	AR least prevalent valvular dysfunction in Tx patients with acute rejection; mild and transient nature because of reabsorption of valvular edema
Tricuspid regurgitation Am J Cardiol. 1994;73:1197-201.	Moderate-severe TR most frequent valvular dysf in Tx ps; 76% TR in Tx pts. w/ right heart failure; due to vavular edema, pulm. hypertens and RV dilatation
Mitral regurgitation Echocardiography. 1992 Mar;9(2):169-74	MR seen in 40% of Tx patients with acute rejection; recommended as indication for immediate biopsy; due to valvular edema
%Ejection fraction Transplant Proc, 38, 636–638 (2006)	Significant decrease in %EF seen in Tx patients with biopsy-proven acute rejection, (50% to 36% during rejection); due to LV systolic dysfunction
Stroke volume Transplant Proc, 38, 636–638 (2006)	Significant decrease in SV seen in Tx patients with biopsy-proven acute rejection (48% to 31% during rejection); due to LV systolic dysfunction
Posterior wall thickness Arq Bras Cardiol. 1989 Sep;53(3):151-5.	PWTh increases associated with acutely rejecting grafts with biopsy grades 3 or 4; due to immune cell infiltration and inflammation
Heart rate variability J Heart Lung Transplant. 1998 Jun;17(6):578-85.	(Low Freq.÷ High Freq - Resp. Peak) freq. domain parameter has a 77% sensitivity, and 76% specificity for acute rejection
Isovolumic relaxation time J Heart Lung Transplant. 2005 Feb;24(2):160-5	Significant decreases (<90ms) in IVRT in biopsy-proven rejecting (grade =>1B) vs. non-rejecting grafts; 53% sensitivity, 68% specificity; due to impaired LV relaxation
Atrial reversal velocity J Heart Lung Transplant. 2005 Feb;24(2):160-5	Significant increases in atrial reversal velocity in biopsy-proven rejecting grafts (grade >=1B) vs. non-rejecting grafts
LV peak systolic, end-diastolic pressure	NOT FOUND IN LITERATURE
Mitral valve half-pressure time Arch Mal Coeur Vaiss. 1997 Dec;90(12):1623-8.	Significant reduction in mitral valve half-pressure time in rejecting vs. non-rejecting but low sensitivity for low %EF values as a result of LV dysfunction
Mitral deceleration time J Heart Lung Transplant. 2005 Feb;24(2):160-5	Significant increases in mitral DT in biopsy-proven rejecting grafts (grade >=1B) vs. non-rejecting grafts
Tricuspid early peak flow velocity Circulation. 1992 Nov;86(5 Suppl):II259-66	Significant increases in tricuspid early filling peak in biopsy-proven rejecting grafts vs. non-rejecting grafts
Tei Index/Index of Myocardial Performance Am J Cardiol. 2002 Sep 1;90(5):517-20	An IMP increase of >/=20% from baseline had 90% sensitivity and 90% specificity in detecting high-grade cardiac allograft rejection
Intraventricular septal thickness Eur Heart J. 1989 May;10(5):400-8	IVS thickness increased significantly in Tx pts. undergoing acute rejection vs. pts. with no rejection (55% incidence in acutely rejecting grafts)
LV end-systolic volume Circulation. 1987 Nov;76(5):998-1008	Increases LVESV in allografts during histological rejection; significant changes compared to prior rejection-free period
Myocardial echogenicity Circulation. 1997 Jan 7;95(1):140-50.	Increases in PWED and septal 2D-integrated backscatter identify mild to severe acute rejection due to myocyte damage
Left ventricular chamber size J Heart Lung Transplant. 1993 Nov-Dec;12(6 Pt 1):1009-17	LV chamber size was significantly decreased in allografts vs. isografts
A-Ar interval (A: mitral late flow; Ar: outflow) J Am Soc Echocardiogr. 1997 Jun;10(5):526-31	Significant decreases in A-Ar interval associated with increasing biopsy scores
Tricuspid E/A ratio J Heart Lung Transplant. 2005 Feb;24(2):160-5	Significant increases in tricuspid E/A ratio in moderate-severely rejecting grafts (grade >=1B) vs. non-rejecting grafts
Early mitral flow velocity (E) Circulation. 1987 Nov;76(5 Pt 2):V86-92	Increasing severity of rejection associated with increased M1 early mitral filling velocity (E-wave)
Pericardial effusion J Heart Lung Transplant. 2005 Feb;24(2):160-5	Pericardial effusion more prevalent in moderately-severely rejecting grafts vs. non-rejecting grafts; due to difference in donor graft size, inflammation
Mitral A-wave duration J Heart Lung Transplant. 2005 Feb;24(2):160-5	Significant decreases in mitral A wave duration in moderate-severely rejecting grafts (grade >=1B) vs. non-rejecting grafts; due to higher LVED pressure
Mitral E/A ratio J Heart Lung Transplant. 2005 Feb;24(2):160-5	Significant increases in mitral E/A ratio (>1.7) in moderate-severely rejecting grafts (grade >=1B) vs. non-rejecting grafts; due to impaired mitral filling and LVEDP
Myocardial blood (coronary) flow Circulation 79:59, 1989	Decreased hyperemic and increased resting MBF in biopsy-proven rejecting grafts vs. non-rejecting grafts; due to endothelial dysfunction, edema, local coagulation
Pulmonary vein systolic flow J Heart Lung Transplant. 2005 Feb;24(2):160-5	Significant increases in pulmonary vein systolic flow in moderate-severely rejecting grafts (grade >=1B) vs. non-rejecting grafts
Superior vena caval systolic flow Circulation. 1992 Nov;86(5 Suppl):II259-66	Significant decrease in SVC flow velocity in rejecting vs. non-rejecting grafts due to lower long-axis shortening of RV assoc. w/ acute rejection; 100% sens, 80% spec
DP/tmax values Circulation. 1989 Jan;79(1):66-75	Significant decrease in DP/tmax values for patients with restrictive-constrictive physiology; associated with higher rejection incidence

14

1.2 THE MOUSE IN TRANSPLANT RESEARCH

1.2.1 Experimental Animal Models of Transplantation

1.2.1.1 Introduction

The basis for much of our current knowledge in physiology, pharmacology, microbiology, immunology, and pathology has been procured from studies in animals (Hau and Van Hoosier, 2003). In transplantation for example, the transition of surgical transplantation from theoretical frameworks to an established and effective therapy for end-stage organ failure has been made possible through the extensive scientific insight gained from experimental animal models (Cramer *et al.*, 1994). Applications to organ transplantation are varied and include kidney, liver, pancreas, skin, bone marrow, small bowel and multi-visceral, and cardiac transplantation; these procedures typically employ rats, rabbits, swine, mice, and primates (Cramer *et al.*, 1994). Historically, the French surgeon Alexis Carrel was the first to develop surgical procedures for experimental transplantation. His inaugural work on establishing vascular procedures for kidney and heart transplantation in dogs was recognized when he was awarded the Nobel prize in 1912 (Hau and Van Hoosier, 2003). Later unsuccessful attempts in clinical organ transplant surgery made in the late 1930's were followed by a renewed interest in the early 1960's due to the production of a novel immunosuppressant, 6-mercaptopurine. The improved survival rates of transplant patients led to increasing numbers of interested clinicians and researchers in the field, and naturally led to the development of sophisticated animal models to investigate basic lines of enquiry in transplantation. These include new pharmacological agents capable of mitigating immune responses towards allografts, effective preservation solutions, elucidating immunological mechanisms of rejection, developing novel sources of grafts (xenografts), and the establishment of post-transplant monitoring protocols (Cramer *et al.*, 1994).

1.2.1.2 Heterotopic Vascularized Cardiac Grafts in Rodents

In experimental cardiac transplantation, rats and mice are the most commonly used animals, though canine, swine, primates and other rodents (*e.g.*, hamsters) have also been employed (Cramer *et al.*, 1994). Vascularized heterotopic heart transplantation is the method of choice in rodent transplantation experiments. Orthotopic heart transplantation is not possible in rodents because cardiopulmonary bypass systems are neither available nor appropriate for the relative small size of these animals. However, it should be noted that since the HHTx rodent model does not possess venous return to the graft, only limited comparisons can be made with the physiology of the orthotopically placed graft.

The HHTx procedure was first described by Abbott *et al.* in 1964 using rats and utilized end-to-end anastomosis of the donor thoracic aorta to the recipient abdominal aorta and donor pulmonary artery to recipient inferior vena cava (IVC) (Abbott *et al.*, 1964b). Abbott's procedure resulted in restricted blood flow to the lower extremities because the recipient vessels were divided below the renal vessels, resulting in paraparesis and paraplegia in the recipient (Cramer *et al.*, 1994). This procedure was later revised by Ono and Lindsey to allow for greater blood flow to the lower extremities (Ono and Lindsey, 1969). They employed end-to-side anastomoses of the donor vessels to the recipient aorta and IVC, resulting in 90% graft and recipient survival (Ono and Lindsey, 1969). Heterotopic cervical transplantation was performed in rats in 1971 (Heron, 1971) and required the anastomosis of the recipient external jugular vein and common carotid artery to the donor aorta and pulmonary artery, though thrombosis occurred in 30% of subjects (Cramer *et al.*, 1994). In 1973, Corry *et al.* were investigating the role of MHC molecules in cardiac graft rejection and concurrently reported performing the HHTx protocol in mice using the same technique established by Ono and Lindsey in the rat (Corry *et al.*, 1973b). Blood flow directionality in this model has been verified using ultrasound biomicroscopy (UBM) in a previous study by our laboratory and is illustrated in **Fig. 1** (Zhou *et al.*, 2007). Briefly, blood from the recipient abdominal aortic circulation flows retrograde into the donor graft ascending aorta, producing a high regurgitant jet, and then diverts into the coronary arteries to perfuse the cardiac graft; the venous blood then drains into the right atrium via the coronary sinus. Blood then fills the right ventricle and is ejected through the pulmonary artery and enters into the recipient IVC (Niimi, 2001). Thus, the model represents a left ventricular bypass with omission of normal pulmonary circulation for blood reoxygenation. Rat HHTx graft blood flow volume comprises about 5% of total blood volume per minute, as assessed by estimations from morphometric measurements of graft dimensions *ex vivo* (Silber, 1979).

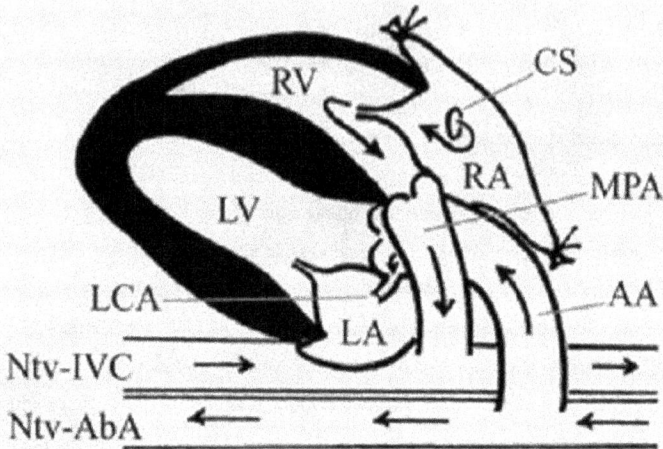

Figure 1 – Schematic Representation of Anastomoses and Blood Flow in Heterotopically Transplanted Cardiac Grafts in Mice. The cardiac graft is heterotopically implanted in the abdomen with the ascending aorta (AA) anastomosed to the native abdominal aorta (Ntv-AbA) and main pulmonary artery (MPA) to the native inferior vena cava (Ntv-IVC). The pulmonary veins and vena cava of the graft are ligated. *CS*, coronary sinus; *LCA*, left coronary artery; *RA*, right atrium; *RV*, right ventricle; *MPA*, main pulmonary artery; *LA*, left atrium; *LV*, left ventricle. Reprinted from *Ultrasound Med Biol,* 33(**6**), Zhou *et al.*, *Morphological and functional evaluation of murine heterotopic cardiac grafts using ultrasound biomicroscopy,* pp. 870-9 (2007) with permission from Elsevier.

1.2.1.3   Vascularized vs. Non-Vascularized Grafts

Primarily or fully vascularized allografts that are perfused by blood from the recipient circulation, such as kidney, liver, and heart, represent many advantages over non-vascularized allografts, especially skin allografts, in investigations evaluating mechanisms of graft susceptibility to rejection and tolerance induction (Jones *et al.,* 2001). It is well documented that non-vascularized grafts are subject to non-specific ischemic degeneration that can lead to inflammation and necrosis, even in syngeneic grafts, which can render these grafts susceptible to subsequent immune rejection (Medawar, 1944). However, the element of vascularity alone has been shown to be insufficient to explain the predisposition for non-

17

vascularized grafts to reject faster than vascularized grafts (Cho *et al.*, 1972). One explanation is the presence of tissue-specific antigens. Work by Steinmuller *et al.* have shown that in hemopoietic chimeras, allogeneic hemopoietic cells survived indefinitely but would not accept skin grafts from the same donor, suggesting that the skin expressed antigens not found on the hemopoietic cells (Steinmuller, 2001). This observation was corroborated by work performed earlier by the same group, which illustrated that Epa-1, a non-H-2 mouse alloantigen defined by MHC-restricted, CD8+ cytotoxic T cells, was a relatively strong determinant of skin allograft rejection but a weaker determinant of heart allograft rejection (Steinmuller *et al.*, 1991). Other important factors to consider are the number and type of antigen presenting cells present in the graft; skin has an abundant number of Langerhans cells – professional antigen presenting cells – which have the capacity to migrate from the graft and efficiently stimulate a large number of alloreactive T cells (Larsen *et al.*, 1990). Finally, the physical size of the graft may also be a determinant in allograft rejection susceptibility; a larger graft or several grafts implanted to the same recipient have a greater total number of cells and thus, require more time for immunological destruction before graft failure ensues (Sun *et al.*, 1996).

It seems apparent that the eventual clinical goals of cardiac transplantation derived from data obtained from experimental models of transplantation are much more readily obtained from models that expressly employ primarily-vascularized cardiac grafts to study immune responses to allograft antigens. Other allografted tissues have the added factors of tissue-specific antigens, types and expression of antigen presenting cells, and physical size of the graft, which carry important ramifications on tolerance induction and conversely, graft rejection (Jones *et al.*, 2001).

1.2.1.4   Advantages of Using the HHTx Murine Model in Transplant Research

The general advantages of utilizing mice in experimental cardiovascular research, such as the induction of transgenic and gene knock-out strains are outlined more thoroughly in **1.3.3.1** and **1.3.3.2**. In transplantation, though the mouse abdominal HHTx is a non-physiologic, non-loaded heart model due to the unique directionality of intra-cardiac blood flow (Bernhard and Konertz, 1983), it is still valuable in investigating a wide variety of transplant-related issues (Cramer *et al.*, 1994; Hau and Van Hoosier, 2003; Konertz *et al.*,

18

1985). The benefits of using the heterotopic abdominal murine model of cardiac transplantation in transplant research include: low surgical costs, limited number of personnel required, availability of a plethora of genetically inbred lines, and isolation and characterization of the MHC molecules in congenic inbred murine lines (Cramer *et al.*, 1994). Studies on MHC molecules form the foundation for understanding immune-mediated responses to allografts. The mouse model also provides important information on graft ischemia-reperfusion injury, rejection pathogenesis, highly specific immunological reagents for detailed examination of the immunopathology of allograft rejection, and the physiologic changes that take place post-transplant in vascularized organs (Cramer *et al.*, 1994). Mice are an order smaller than rats and therefore occupy less space, are cheaper, and are easier to maintain, which gives them excellent value when designing experiments.

1.2.2 Experimental Applications of the HHTx Mouse Model

1.2.2.1 Organ Preservation

Rodent heterotopic models of transplantation have been previously used to investigate the efficacy of novel preservation (Makowka *et al.*, 1989) and cold cardioplegia solution (McGregor *et al.*, 1984) as well as to determine the extent of myocardial injury subsequent to hyperkalemic cardioplegia and prolonged hypothermic ischemia (McGregor *et al.*, 1983). Post-transplant graft status, enzymatic markers of tissue damage and histologic evaluation are commonly used to clinically assess graft viability and these parameters can be tested in HHTx rodent models. McGregor *et al.* have demonstrated, for example, the beneficial effects of St. Thomas' cardioplegia solution in augmenting myocardial protection, as assessed by serum creatinine kinase activity (McGregor *et al.*, 1984). The University of Wisconsin solution (UW lactobionate) has also been investigated in HHTx rodent models to determine the degree of cardiac injury, as determined by biochemical markers and histology, in comparison to traditional solutions (Makowka *et al.*, 1989). These studies often provide critical information for planning larger animal studies (Cramer *et al.*, 1994).

1.2.2.2 Graft Surveillance and Rejection

HHTx rat and mouse models are widely applied for immunologic and histological studies (West *et al.*, 1995; Akashi *et al.*, 2005; Fan *et al.*, 2005; Kosuge *et al.*, 2006). Investigation of effector mechanisms of graft-infiltrating cells by utilizing rat and mouse

19

models of adoptive transfer followed by HHTx surgery has led to a wealth of information regarding immunopathological mechanisms; a sample of the literature includes:

i) Understanding the important role of the graft and CD4+ regulation in inducing acceptance of cardiac allografts (Hofmann *et al.*, 2004);

ii) Determining the critical role of CD4+ and CD8+ T lymphocytes in stimulating graft rejection and their mechanistic roles (Honjo *et al.*, 2004; Oluwole *et al.*, 2003; Grazia *et al.*, 2004);

iii) Investigating the primary role for delayed-type hypersensitivity in vascularized organ allograft rejection (Lowry *et al.*, 1983; Burdick and Clow, 1986);

iv) Evaluating the pivotal mechanisms behind the interaction of host CD4 T cells with donor MHC II molecules in mediating rejection of murine cardiac grafts (Campos *et al.*, 1995; Gould and Auchincloss, Jr., 1999; Corry *et al.*, 1973b; Ogawa *et al.*, 2004);

v) The potential therapeutic effect of antagonism of chemokine (CCR5) and chemokine receptors (CXCR3) in inhibiting both acute and chronic allograft rejection (Akashi *et al.*, 2005);

vi) Understanding the association of increased Fas ligand mRNA expression with myocardial apoptosis and ischemia-reperfusion injury and the eventual progression to chronic rejection (White *et al.*, 1997); and,

vii) Studying the post-transplant shift in expression of genes classified as defense, communication, and metabolism during the graft response to transplantation injury and rejection (Christopher *et al.*, 2003).

1.2.2.3   *Pharmacological Testing and Immunosuppression*

A number of pharmacological agents have been tested in rodent HHTx models for safety and efficacy prior to studies in larger animals and validation to clinical trials. A few examples include Cyclosporin A (Wang *et al.*, 2003), 15-deoxyspergualin (Yuh and Morris, 1993), Brequinar sodium (Cramer *et al.*, 1992b), pioglitazone (Kosuge *et al.*, 2006), sirolimus/rapamycin (Koehl *et al.*, 2004), mycophenolate mofetil (Shimmura *et al.*, 2006), and FTY720 (Nikolova *et al.*, 2001). Other studies, such as Mitsuhashi *et al.*'s, have used gene therapy to induce xenograft tolerance and provided the first demonstration of permanent survival of alphaGal+ hearts following transplantation with autologous bone marrow transduced with porcine GalT-expressing lentivirus (Mitsuhashi *et al.*, 2006). Co-stimulatory blockade antibodies (Kosuge *et al.*, 2003), genetically engineered dentritic cells (Bonham *et al.*, 2002) and other agents (Nguyen *et al.*, 2006) to induce cardiac allograft tolerance in mouse and rat HHTx models. Several monoclonal antibodies targeting molecular epitopes expressed on vascular endothelium and activated T cells have been utilized to induce tolerance and/or improve long-term graft viability and have included intercellular adhesion molecule-1 (ICAM-1) (Isobe and Ihara, 1993), lymphocyte function associated antigen (LFA)-1 and vascular cell adhesion molecule (VCAM-1) (Suzuki *et al.*, 1999).

1.2.2.4   Xenotransplantation

Cordant and discordant xenograft rejection have been studied in several common species combinations, such as mouse->rat, hamster->rat, guinea pig->rat and hamster->guinea pig, to examine the efficacy of immunosuppressive drugs (Cramer *et al.*, 1992a), total lymphoid irradiation, antibodies (Steinbruchel *et al.*, 1991), and inflammatory mediators (Filipponi *et al.*, 1989) in settings of accelerated and hyperacute rejection.

1.2.3   Assessing HHTx Graft Rejection in the Mouse and Rat
1.2.3.1 Palpation and Histology

Two common methods for assessing graft function in rat and mouse HHTx models employ finger palpation of cardiac contractions (Corry *et al.*, 1973b; Corry *et al.*, 1973a) and histology (Abbott *et al.*, 1964a; Abbott *et al.*, 1965). The evaluation of graft viability by finger palpation has traditionally been quantified by a subjective scale where an observer scores the force of contraction from '0' (cessation of graft contraction) to '4' (vigorous graft contraction). Several criticisms of this method of assessment include its subjectivity and therefore a potentially large inter-observer variability. To date, no systematic study has validated this approach as a reliable and accurate method of graft assessment. Furthermore, this method is insensitive for detecting early changes PTx and fares poorly in the evaluation of long- term graft survival (Mottram *et al.*, 1988). This precarious approach is also affected by the ubiquitous presence of left ventricular thrombus seen in iso- and allografts (Zhou *et al.*, 2007), which may potentially 'dampen' the force of ventricular contraction felt and

21

falsely give the impression of decreased graft viability. The fate of LV cavity thrombi and the rate of formation, shrinkage, degradation, etc. have not been previously described and may have important ramifications on finger evaluations of the intensity of pulsation. Finally, one of the many difficulties of using finger palpations is the observation of faint ventricular pulsations that may occur even after advanced acute graft rejection; this effect is commonly due to recipient systemic circulation within the graft (Cramer *et al.*, 1994).

Most investigators, however, employ histological specimens for evidence of cellular infiltrate and myocyte damage either as an adjunct or as stand-alone method of assessing graft rejection. The 'gold standard' for diagnosing graft rejection in clinical surveillance and experimental research remains histology (Caves *et al.*, 1973); (Affleck *et al.*, 2001; Galinanes and Hearse, 1991). Biopsies however are limited by their invasive nature which may cause injury to cardiac structures and are not feasible in mice and rodents due to the small size of the rodent heart (Zhou *et al.*, 2007). Furthermore, reports of patchy sites of cellular infiltration and tissue necrosis have often questioned the reliability of biopsy findings, even when several tissue sites are selected (Weller *et al.*, 2003; Scherrer-Crosbie *et al.*, 2002; Nakhleh *et al.*, 1992). Biopsies are also not amenable to the possibility for serial measurements to be made in the same animal over time, thus limiting their use for short- and long-term monitoring studies.

1.2.3.2 Electrocardiography (ECG)

Electrocardiography (ECG) recordings have been performed in anaesthetized animals with needle electrodes inserted subcutaneously in the lower abdomen and limbs of the rat (Abbott *et al.*, 1965) and mouse (Superina *et al.*, 1986). Early attempts reported an accurate determination of an episode of graft rejection using ECG, though high graft heart rates (350-450 BPM) and poorly defined S-T segments have precluded consistent interpretation of ECG spectra in the rat HHTx model (Bishop S., 1980). Cardiac abnormalities such as bradycardia and arrhythmias have been reported in ECG spectra several hours PTx, though normal waveforms are usually seen in the next few days in allografts; in isografts, consistent readings can be obtained for as long as 1-yr. post-implantation (Heron, 1972). The major limitation of using ECG electrodes inserted abdominally is the possibility that the ECG tracing can be affected by the location of electrodes. This complication is compounded by the lateral and horizontal movements of the graft within the abdomen (Mottram *et al.*, 1988) and interference from surrounding structures such as the intestines and peri-graft effusion (Zhou *et al.*, 2007). More recently, analysis of heart rate variability (HRV), specifically total power of HRV, was found to be significantly higher in histologically-confirmed acutely rejecting rat HHTx grafts when compared to isografts (Wada *et al.*, 1999). Significantly decreased peak-to-peak amplitudes of the QRS complex and heart rate on day 5.5 PTx or later were also found in rejecting grafts but these observations were absent in isografts. However, a telemetric ECG radiotransmitter must be implanted near the graft and specialized highly technical equipment is required. In mouse HHTx studies, allografts were rejected in 10-14 days and showed a rapid fall in palpated heart rate, measured heart rate, and ventricular QRS complex voltage (Mottram *et al.*, 1988).

1.2.3.3 Nuclear Magnetic Resonance (NMR) and Related Applications

Studies utilizing NMR imaging have investigated the potential of this modality to detect changes in myocardial metabolism during acute rejection in rat HHTx grafts as a way of validating this approach for clinical use in transplant patients (Haug *et al.*, 1987). Haug *et al.* found that grafts with histological rejection showed relatively low levels of phosphocreatinine and high levels of inorganic phosphate versus isograft controls (Haug *et al.*, 1987). More recent efforts however utilize the heralded medical imaging applications of NMR technology, or magnetic resonance imaging (MRI). One intriguing study by Kanno *et al.* reported that rat HHTx allografts displayed a significant reduction in MR signal intensity after injection of dextran-coated ultrasmall superparamagnetic iron oxide (USPIO) particles, whereas no differences were detected in isografts. The reduction was confirmed to be due to macrophage uptake of these particles in acutely rejecting allografts and the reduction in intensity was reversible with immunosuppression (Kanno *et al.*, 2001). Investigations to improve the sensitivity of MRI have included using novel contrast agents (Penno *et al.*, 2006) and *in vivo* cell tracking of individual macrophages (Wu *et al.*, 2006), and T cells (Dodd *et al.*, 1999), which have been reported to uptake the contrast agent *ex vivo* and in turn, allow them to be monitored during cellular infiltration in the rejection process as increased MR signals in localized regions of the graft. Other MR-based approaches have been investigated to assess morphological and physiologic changes in rat HHTx grafts, such

23

as degree of edema, ventricular mass and volume changes, SV, %EF, and LV chamber dimensions, but have not yet been validated for wide-scale experimental use (Wu *et al.*, 2004). NMR-based approaches have several limitations, however, which include the need for highly technical protocols, very costly equipment, its invasiveness, and the inherent inability in obtaining serial data in the same animal subject in perfusion studies. MRI is also a less desirable option for mice, whose haemodynamic status can be significantly altered by the imaging procedure and its utilization of deep or prolonged anaesthesia (Yang *et al.*, 1999).

1.2.3.4 Molecular and Biochemical Markers of Cardiac Graft Rejection

Distinct cytokine profiles expressed during the rejection process have been shown to play an essential role in regulating the pattern of rejection and outcome of immunosuppressive therapy (Wang *et al.*, 2003). Investigators have proposed surveillance of the expression of transcripts, chemical messengers and key molecules expressed on dysfunctional endothelium as a non-invasive and sensitive method of evaluating the degree of rejection. There is a large body of work focused on the potential for scintigraphy to detect cardiac allograft rejection non-invasively in mouse and rat HHTx models. Previous studies have employed various isotopes and targets to study the effects of acute rejection on radiotracer uptake. These include [123]iodine-labeled and ^{111}In anti-ICAM-1 mAbs (Isobe *et al.*, 1993; Ohtani *et al.*, 1995), ^{111}In-labeled anti-MHC class II antigen mAbs (Isobe *et al.*, 1992), and ^{111}In-antimyosin mAbs (Isobe *et al.*, 1991), and most have documented increased accumulation of the respective isotope in association with histologic rejection. The requirements for costly antibodies and radiological materials as well as the lack of consensus on specific molecular targets important for detecting rejection however have precluded these approaches for routine use. Work by Suzuki *et al.* and others have focused on assessing upregulation of cytokines (*e.g.*, IL-2, IL-4, IL-10, IL-12) and their transcripts (*e.g.*, IFN mRNA) as assessed by *in situ* RT-PCR and immunohistochemistry in infiltrating cells during rejection. They reported that this provides a sensitive method for the non-invasive detection of murine graft rejection (Suzuki *et al.*, 1998; Wang *et al.*, 2003). Nonetheless, there does not seem to be a consensus on a definitive, single test for rejection and the majority of studies suggest that experimental results should used be as an adjunct to histology. Cellular tracking

24

accomplished by *in vivo* bioluminescence imaging, however, has been performed to visualize migration and proliferation of CD5+ passenger leukocytes in both isograft and allograft murine recipients and appears to be a promising diagnostic test for graft rejection (Tanaka *et al.*, 2005d).

1.2.3.5 Ultrasound and Related Applications

The association of backscatter analysis and levels of apoptosis as a result of immunopathological responses to allografts is reviewed in **1.3.3.** The current state of the applications of conventional echocardiography and the novel UBM modality to image acutely rejecting HHTx cardiac grafts non-invasively is summarized in **1.3.5.** Other notable applications, including 3-D echocardiography (Dawson *et al.*, 2004) and Tissue Doppler imaging (Sebag, 2005) have not, as far as we know, been applied to investigate ventricular global and regional function in rodent HHTx grafts, perhaps because the former method requires a large amount of off-line computer and mathematical analysis, and the latter gives little information on cardiac morphology. The recent development of gas-filled, lipid-based antibody-conjugated microbubbles for binding cellular and biochemical targets implicated in the pathogenesis of graft rejection has proposed myocardial contrast echocardiography as an exciting and promising method of detecting rejection in experimental and potentially, clinical settings (Lindner, 2004). Recently, ultrasound imaging of acute cardiac transplant rejection in rat HHTx grafts following injection of microbubbles targeted to ICAM-1 illustrated the ability of adhered bubbles to induce significantly increased myocardial video-intensity in rejecting versus control grafts (Weller *et al.*, 2003). Similar approaches have used leukocyte-targeted bubbles to assess non-invasively the magnitude of intramyocardial infiltration of macrophages and T lymphocytes (Lindner *et al.*, 2000). Microbubbles are currently being tested for UBM applications (Goertz *et al.*, 2005).

1.3   CARDIAC ULTRASOUND IMAGING OF THE MOUSE

1.3.1   *Development of Echocardiography*

Though ultrasound has always existed in nature, it wasn't until the discovery of piezoelectricity in 1880 that scientists began to appreciate the potential of this technological discovery (Curie P and Curie J, 1880). The first investigation to examine the heart ultrasonically was reported by Wild *et al.* in 1957, though applications were restricted to autopsy specimens (Wild and Neal, 1951). Keidel, the first physician to use ultrasound to image the heart, reported that he could obtain acoustic shadows when an ultrasonic beam was directed to the chest (Keidel WD, 1950). However, the discovery of contemporary echocardiography is typically accredited to Edler and Hertz in 1954 (Edler and Hertz, 1954), who, among other things, were investigating the application of ultrasound in diagnosing mitral stenosis (Edler I, 1956). Their work also prompted the use of time-motion or M-mode imaging. Later European researchers were able to quantify left atrial masses using cardiac ultrasound (Effert and Domanig, 1959) and Japanese-led efforts discovered and developed Doppler ultrasound for clinical uses (Yoshida *et al.,* 1956).

1.3.2 *Ultrasound Biomicroscopy (UBM)*

1.3.2.1 Introduction

Ultrasound biomicroscopy (UBM) is a relatively new imaging technique that uses high-frequency ultrasound to produce images of small biological structures at near microscopic resolution. This technique was developed in Toronto, Canada based on the basic research conducted in the labs of F. Stuart Foster. UBM instruments are now in world-wide distribution (Foster *et al.,* 1993; Foster *et al.,* 2000).

1.3.2.2 Advent of UBM

In spite of the growing popularity of ultrasound in the 50s through the 70s, it was several decades earlier, in 1935, when Sokolov introduced an application of ultrasound that would change biomedical imaging forever – its ability to image cellular microstructures using an acoustic microscope (Sokolov, 1935). Sokolov's discovery eventually led to the development of the first scanning laser acoustic microscopes (SLAMs) (Kessler *et al.,* 1972) and scanning acoustic microscopes (SAMs) (Lemons and Quate C.F., 1974) in the 1970s. In the mid-1980's, investigations were conducted into imaging tumour spheroids using pulse-echo imaging systems operating at frequencies an order of magnitude higher (Sherar *et al.,* 1989) and the potential of non-invasive, serial imaging of subsurface tissue was soon realized (Foster *et al.,* 2000).

26

1.3.2.3 Physics of UBM Imaging

UBM is essentially an extension of the B-mode backscatter methods developed for clinical imaging in the 3 to 10-MHz frequency range. Acquiring UBM images of microscopic, subsurface structures is dependent on the transducer. UBM transducers can have a frequency range of 20-50 MHz, meaning that a radio-frequency pulse of 20-50 MHz is produced by the piezoelectric crystal of the transducer (Sherar and Foster, 1989). This radiofrequency travels through the body tissue and is reflected back to the transducer. The reflected radio-frequency is processed by a signal processing unit and determines three important parameters of the received information: directionality, strength, and time elapsed. In the interest of preserving image quality, the transducer typically has an open crystal (Foster *et al.*, 1993).

The most important considerations for imaging performance of any ultrasound scanner are frequency, geometry of transducer, and tissue properties in accordance with the laws of diffraction (Foster *et al.*, 2000). High-frequency ranges typically have trade-offs in axial and lateral resolution as well as decreased depth of field; therefore, tissue properties - primarily attenuation - must be considered when determining the ideal high-frequency ranges to be used. Attenuation in ultrasound is the reduction in amplitude of the ultrasound beam as a function of distance through the imaging medium. Accounting for attenuation effects in ultrasound is important because reduced signal amplitude can affect the quality of the image produced (Bushong and Benjamin, 1991). Structural tissues, such as skin, arterial walls, and sclera tend to have the highest attenuation coefficients, whereas the iris and cornea have typically low losses, approaching that of water (Foster *et al.*, 2000). Differences in the concentration and organization of structural proteins, like collagen for example, have been suggested as the possible reason for the attenuation disparities in tissues. The degree of backscatter is also an important consideration and, like attenuation coefficients, is dependent on the tissue type (Foster *et al.*, 2000).

1.3.2.4 UBM Modes and Functions

Because of its high resolution, the UBM has been used to image small animals, predominantly the mouse, non-invasively in biological research. The spatial resolution of a two-dimensional image is up to ~50 μm, with penetration depth of ~20 mm. The basic functional modalities of UBM include: *B-mode,* which provides two-dimensional images of the heart and vessels, abdominal organs, skin, eye and tumor for morphological and dimensional observation in mice; *M-mode,* which yields the dynamic change of position of moving structures, such as the walls of the left ventricle and arteries over time; cardiac systolic function and arterial dimensions can be measured using this functionality; *Doppler mode,* which measures the blood flow velocity at the vessel or orifice of interest; *3D reconstruction,* which can acquire a three-dimensional volume data set of a target structure, measure tissue volume, and visualize the surface image of internal structures; and *EKV* (ECG based Kilo-Hertz Visualization) technology, which retrospectively recreates a 2-D cardiac image equivalent to that acquired at a frame rate of 1000 Hz; it is suitable for evaluating LV wall motion and valvular movement (Zhou *et al.,* 2004). Other newer functions include *anatomic M-Mode,* which allows the sonographer to acquire an time course M-Modes in the anatomically correct plane if, for example, a longitudinal axis view of the LV cavity cannot be acquired perpendicularly with the cursor in conventional M-mode. Finally, *power-Doppler* is a relatively new mode available on UBM and encodes the power in the Doppler signal in color. This parameter is fundamentally different from the mean frequency shift, which is determined by the velocity of the red blood cells, while the power depends on the amount of blood present.

1.3.2.5   Clinical UBM

Clinical applications of ultrasound biomicroscopy (UBM) are varied and reflect the emerging influence of this novel imaging modality, not only in experimental spheres, but those of importance to routine clinical procedures. Three main fields with varying technical specifications and frequency ranges have been established for UBM clinical applications: ophthalmology in the 40-60-MHz range (Pavlin *et al.,* 1990; Sherar *et al.,* 1989), dermatology in the 20-MHz range (Hoffmann *et al.,* 1989; Hoffmann *et al.,* 1990) and intravascular ultrasound (Bom *et al.,* 1989; Meyer *et al.,* 1988; Nissen *et al.,* 1990; Yock *et al.,* 1989). Several examples of applications in the field of ocular UBM imaging include: segment tumours, glaucoma, scleral disease, corneal disease, intraocular lens assessment and trauma (Pavlin CJ and Foster FS, 1995). Because of its ability to image microscopic structures of critical importance, such as the cornea, iris, ciliary muscle and lens, UBM imaging has contributed much to the accurate diagnosis of several forms of glaucoma, in visualizing the progression of scleral disease, and in evaluating blood flow in the microcirculation of the eye (Foster *et al.*, 2000). Dermatological imaging offers a high-resolution, non-invasive assessment of skin elasticity (Berson *et al.*, 1992), skin thickness, (Foster *et al.*, 2000) and skin tumour dimensions and depth (Harland *et al.*, 1993). Intravascular imaging, also known as intravascular ultrasound, has been preferred over angiography to study plaque build-up, calcium content, tissue trauma, and lumen dimensions (Mintz *et al.*, 1996). Other non-vascular applications include visualization of the urinary tract, Fallopian tube, esophagus, bowels, and bronchial tree (Goldberg, 1991). Finally, cartilage imaging has also been investigated in the detection of osteoarthritis, though this work has only been validated in animal models (Spriet *et al.*, 2005).

1.3.2.6 Biological Applications of UBM

Aside from adult mouse cardiovascular imaging which will be discussed below, UBM has become an established tool for the study of developmental processes as well as the genetic events that give rise to varying phenotypes during embryogenesis. For example, UBM imaging has been used to assess the critical role of Wnt-1 in mid hindbrain development using Wnt-1 knock-out mice (Turnbull *et al.*, 1995), the requirement for endothelial NO synthase expression in meeting increased cardiovascular demands during pregnancy (Kulandavelu *et al.*, 2006), and the rapid onset of diastolic dysfunction in embryos lacking nuclear factor of activated T-cells (NFATc1) gene (Phoon *et al.*, 2004). Tumour growth has also been investigated in mice with cutaneous melanomas; UBM-derived tumour dimensions and volumes showed good correlation with histopathology and low relative error (Turnbull *et al.*, 1996).

1.3.3 Histopathological Changes and Ultrasound Backscatter

1.3.3.1 Ultrasound Backscatter and Apoptosis

Recent studies associating the presence of apoptotic cells and the apparent increase in backscatter or visual intensity during echocardiographic imaging have given sonographers and those interested in cell death an appreciation for the diagnostic capabilities of ultrasound to detect dead tissue, primarily via apoptosis, non-invasively (Czarnota *et al.*, 1999).

Increased end-diastolic posterior wall and septal 2D-integrated backscatter in serial studies has been shown to permit reliable identification not only of moderate and severe, but also of mild acute rejection (Angermann *et al.*, 1997). The mechanism behind the increased ultrasonic signal was verified experimentally to be due to subcellular nuclear changes and condensation, followed by cellular fragmentation that occurs during apoptosis (Czarnota *et al.*, 1997). Using high-frequency (40 MHz) ultrasound (UBM) imaging, apoptotic cells were found to have a significantly higher ultrasound scattering efficiency (25- to 50-fold change in intensity) over normal cells. The result is that areas of tissue undergoing apoptosis become much brighter in comparison to surrounding viable tissues (Czarnota *et al.*, 1999). In cervical HHTx goat models, an association of the degree of brightness from serial 2-D myocardial echo images with myocardial edema as a result of biopsy-proven hyperacute rejection was suggested almost two decades ago (Chandrasekaran *et al.*, 1987), though the direct causes have not been studied. A corroborating study by Hete *et al.* verified that integrated backscatter increased 1-2 days PTx in the same HHTx goat model, but decreased thereafter (Hete *et al.*, 1990). Other studies in HHTx dog grafts showed that acute cardiac rejection is associated with a progressive increase in mean gray level (a measure of transmitted light energy) (Stempfle *et al.*, 1993). These findings, coupled to the observations made by Czarnota's group, have partially explained the observation of increased echo signal as a result of cellular changes induced by apoptosis. These findings seem to have relevance in our collaborative study, in which we identified a progressive increase in backscatter in murine HHTx isografts post-transplant using UBM (Zhou *et al.*, 2007). The observed increases in backscatter were not quantified in these mice and only isografts were studied. To our knowledge, no study has investigated a possible association between apoptosis and/or cellular changes as a result of acute cardiac rejection with augmented ultrasonic image intensity in rodent cardiac HHTx allografts.

In transplantation, there are two primary causes of apoptosis (programmed cell death): ischemia-reperfusion injury and immune cell-mediated apoptosis. Reports of increased backscatter during ultrasound imaging of clinical and experimental grafts have associated cellular changes as a result of histopathological mechanisms due to apoptosis as a possible cause. The mechanisms for causing apoptosis are briefly described below as well as

30

the possible association investigators have proposed in support of the association of apoptotic cell death with ultrasonic changes.

1.3.3.2 Ischemia-Reperfusion Injury and Apoptosis

Ischemia-reperfusion injury is a ubiquitous problem for organ preservation and cardiac transplantation (Hicks *et al.*, 2006) and is an important element worth describing for understanding early (and perhaps long-term) changes that occur in murine HHTx. Reperfusion of the graft following the ischemic time – the period during which blood flow is abrogated in order to explant and implant a donor organ – has been established as a cause of injury to the organ. The injury is mediated by a multi-factorial chain of events that involves depletion of energy substrates, alteration of ionic homeostasis, production of reactive oxygen species, and cell death by apoptosis and necrosis (Jassem *et al.*, 2002). Molecular changes that result in apoptosis, for example, have been shown to induce organ damage and are implicated in primary graft failure (Hicks *et al.*, 2006; Tanaka *et al.*, 2005a). One such mechanism involves mitochondrial permeability transition pore (MPTP) opening in initiating caspase enzyme-mediated apoptosis of organ tissue during reperfusion (Jassem *et al.*, 2002). Similarly, in rat HHTx models, Raisky *et al.* have shown that MPTP may play a critical role in cardiomyocyte apoptosis associated with acute cardiac graft rejection in rat HHTx grafts (Raisky *et al.*, 2004) and corroborating studies have shown that prolonged cold ischemia in rat cardiac allografts promotes ischemia-reperfusion injury (Tanaka *et al.*, 2005c).

1.3.3.3 Immune-mediated Cellular Apoptosis

A separate cause of apoptosis of graft tissue is via the effector mechanisms of cytotoxic T cells (CTLs), which are well documented to be of critical importance in the pathogenesis of acute cardiac rejection (Bergese *et al.*, 1997). Interestingly, possible methods for the cytotoxic action of T cells during the destructive phase of acute allograft rejection include the TNF/TNF receptor system (Higuchi and Aggarwal, 1994), the Fas/Fas ligand system (Ju *et al.*, 1994), and the granzyme/perforin system (Griffiths and Mueller, 1991), with the latter mechanism found to be also correlated with histological rejection score and a decrease in diastolic function in clinical cardiac transplant recipients (Alpert *et al.*, 1995). In animal studies, Bergese *et al.* discovered that murine HHTx cardiac isografts display no

31

detectable terminal deoxynucleotidyl transferase biotin-dUTP nick end labelling-positive (TUNEL+) cells, whereas rejecting cardiac allografts display diffuse TUNEL+ cells, peaking on day 3 and declining to 50% of peak by the day of rejection (approximately day 10). TUNEL+ cells were also localized to regions of cellular infiltrate rather than myocyte regions (Bergese *et al.*, 1997). Taken together, it is apparent that immune cell-mediated apoptosis during acute rejection is important for graft function.

1.3.3.4 *Ultrasound Backscatter and Necrosis*

There is evidence that another form of cell death, namely necrosis, can result in increases in ultrasound backscatter. Czarnota *et al.* reported that increased backscatter signal was detected in heat-killed human acute myeloid leukaemia cells, though the change in signal intensity was not as large as that caused by apoptotic cells (Czarnota *et al.*, 1997). Immune-mediated destruction of cardiac graft tissue and necrosis of graft parenchyma of acutely rejecting murine HHTx grafts have been well documented and have been associated with the action of cytotoxic T cells and infiltrating polymorphonuclear cells due to ischemia, respectively (El Sawy *et al.*, 2004; Miura *et al.*, 2003). Similar to apoptosis, ischemia-reperfusion of the graft has also been established to result in necrosis (Jassem *et al.*, 2002). Previously, Smith *et al.* reported that long operative ischemic times (~120 minutes) induced 30% myocardial necrosis or scarring of myocardial tissue in HHTx murine isografts (Smith *et al.*, 1987). Necrosis may therefore also represent an important mechanism for the apparent increased echo scores.

1.3.4 **Cardiovascular Phenotyping of the Mouse Heart**

1.3.4.1 *Advantages of Murine Ultrasound in Cardiovascular Research*

There is currently much vested interest in developing and utilizing existing ultrasound technology to study the biochemical and genetic causes of a plethora of cardiovascular diseases and conditions. Because of its non-invasive nature, high-resolution, ability for serial imaging, and the tremendous wealth of information that can be gained, echocardiography is a popular choice for many investigators who wish to study animal models of cardiovascular disease (Zhou *et al.*, 2004). The availability of transgenic and knock-out mouse models, well characterized strains, 90% genome similarity to humans, short gestation periods, low maintenance costs and small size (20-40 grams), have made the mouse a popular choice for studying the development of congenital and progressive cardiomyopathies (Scherrer-Crosbie, 2006). Other techniques for phenotyping murine models of cardiovascular disease, such as magnetic resonance imaging (MRI) and invasive haemodynamics, are less desirable options for mice (Yang *et al.*, 1999). Alternative approaches, such as ventricular catheterization (Lorenz and Robbins, 1997), radiolabeled microspheres, and thermodilution techniques (Barbee *et al.*, 1992) require instrumentation and microsurgical techniques and are difficult to perform more than once in the same animal (Yang *et al.*, 1999).

1.3.4.2   *Mouse Models in Experimental Cardiology*

Prior to the 1970's, much of the scientific progress attained in cardiovascular research was conducted on large animals, such as rabbits and pigs. The recombinant DNA technology that began flourishing in the same decade, coupled to the gradual realization of the relative ease of inducing transgenic changes in mice, promulgated the laboratory mouse as the most common animal model to study cardiovascular disease (Madamanchi, 2004).

There are several research paradigms based on the mouse in cardiology. Mouse models of atherosclerosis are among the most prominent applications. Since wild-type mice are innately resistant to the formation of arterial lesions and hyperlipidemia normally characteristic of humans with cardiovascular disease, knock-out models, which employ targeted loss of function mutations of specific genes, have been produced. In apolipoprotein E (apoE) -/- KO mice for example, all subjects show atheromatous streaks and a lipid profile consistent with cardiovascular disease pathology (Zhang *et al.*, 1992). Further genetic manipulations can be performed against this genetic background to elucidate the role of single genes in accelerating disease progression (van den Maagdenberg *et al.*, 1993). Other popular models include low-density lipoprotein receptor (LDL) -/- knock-outs (Yokode *et al.*, 1990) and knock-in models, which have been used to replace endogenous ApoE alleles with mutant or variant analogues (Sullivan *et al.*, 1998).

A sample of mouse models of relevance in cardiovascular research include: vascular remodelling, which is focused on smooth muscle responses to mechanical arterial injury and novel drug-eluting stents (Lindner *et al.*, 1993); hypertrophy and the role of contractile

33

protein abnormalities (Geisterfer-Lowrance *et al.*, 1996); myocardial ischemia-reperfusion injury subsequent to coronary artery occlusion, and the molecular cascades responsible for myocyte destruction (Metzler *et al.*, 2001); and angiogenesis induced by limb ischemia from permanent ligation of the femoral artery and monitoring of new vessel growth (Emanueli and Madeddu, 2001).

1.3.4.3   Mouse Echocardiography

There is an overwhelming body of literature on the use of ultrasound – both with conventional clinical echo or high frequency (UBM) systems – to assess phenotypically cardiovascular function in developing and adult mice, and any attempt to summarize thoroughly these results is impractical. Nevertheless, most investigations utilize common parameters established by clinical echocardiography and in some cases, have been technically modified and adapted for murine imaging. These parameters of cardiovascular function may be generally categorized as measures of cardiac morphology, blood flow, cardiac systolic and diastolic function, and vascular impedance, with some overlap with other parameters in other categories (Phoon and Turnbull, 2003). A brief description of the historical obstacles in murine echocardiography as well as general anaesthesia will be addressed before discussing the applications of the above imaging applications in mice of wild-type and disease states.

1.3.4.4   Previous Problems

Utilization of ultrasound-based technology for the non-invasive characterization of a large number of cardiovascular phenotypes in mice is on the rise (Scherrer-Crosbie, 2006). A previous obstacle to the application of echocardiography to image the mouse heart was the low frame rate precluding acquisition of accurate information in the face of a rapid mouse heart rate (>500BPM) and second, low transducer frequencies resulting in poor resolution images. These obstacles have been addressed by the development of newer, high-frequency transthoracic transducers, employing novel broadband annular phased-array and linear probes with small footprints (imaging fields). The results are increased lateral resolution, high frame-rate and improved near-field imaging, allowing for the generation of high-quality images of the mouse heart (Collins *et al.*, 2003).

1.3.4.5   Anaesthesia

Other developments in the field of murine echocardiography include the variety of anaesthetics available, which allow for image acquisition to be acquired easily in the much slower beating heart (~50% reduction), as well as the relative ease in handling unconscious animals. However, several studies have shown that certain anaesthetics may have significant effects on cardiovascular function (Collins *et al.,* 2003). Several procedures are available in mouse anaesthesia and include intraperitoneal injection of ketamine:xylazine mixture, tribromoethanol (Avertin), chloral hydrate, or barbiturates, as well as inhalation of isoflurane or halothane. Generally, most anaesthetics and inhalants cause cardiovascular and respiratory depression. Ketamine has been alleged to have less of a cardiodepressive effect when administered with xylazine or diazepam, though the mixture can result in hypothermia and negative inotropic and chronotropic effects (Hart *et al.,* 2001; Arras *et al.,* 2001). Moreover, Roth *et al.* have reported that intraperitoneal agents tribromoethanol, ketamine-midazolam, or ketamine:xylazine induce early cardiodepression and decreased shortening fraction over the course of a 20-min echocardiographic study, although tribromoethanol had a lesser effect on cardiac haemodynamics (Roth *et al.,* 2002). Inhaled isoflurane gas anesthesia, however, resulted in the most stable fractional shortening and end-diastolic dimensions and the most reproducible measurements in repeated studies (Roth *et al.,* 2002). The general agreement is that the choice of anesthetics, dosing regimen, and timing of data acquisition must be tailored to the specific strain, gender, model and mutation when performing echocardiographic studies in mice (Collins *et al.,* 2003).

Investigators concerned with the alteration in cardiovascular function induced by anaesthesia have opted to image conscious mice; this method however may require multiple training sessions in handling the animals and is not possible in catheterized or instrumented animals (Yang *et al.,* 1999). It is also possible that the stress of restraint can affect cardiac function in these animals.

1.3.4.6   Cardiac Dimensions and Morphology

Two-dimensional echocardiography has been previously employed to quantify non-invasively ventricular wall thicknesses, chamber dimensions, left atrial size, aortic root and main pulmonary artery dimensions, mitral valve, and valvular structures, of which some may be obtained by 2-D B-mode images or using M-mode analyses (Collins *et al.*, 2003). Parasternal long-axis views can be obtained by orienting the tip of the transducer to the left shoulder of a mouse laying supine, which allows for the general assessment of overall LV size and function. A 90° clockwise turn towards the right shoulder yields the short-axis view, which permits visualization of the LV anterior and posterior walls, interventricular septum, papillary muscles, and LV cavity (Scherrer-Crosbie, 2006). Ventricular end-systolic and end-diastolic diameter measurements can be made with less than 20% intra- and interobserver variability in both B-mode and M-mode, though obtaining wall thicknesses tend to give higher variabilities (Scherrer-Crosbie *et al.*, 1999; Yang *et al.*, 1999). In contrast to conventional, lower-frequency systems which cannot reproducibly image the right ventricle nor obtain apical 2 and 4-chamber views in rodents, the UBM provides high-resolution and consistent image acquisition of the right atrium and ventricle in a longitudinal section from a left parasternal window. This is in addition to the excellent procurement of LV chamber and valvular orifice dimensions and morphology, ventricular wall and septal thicknesses in both B-mode and M-mode (Zhou *et al.*, 2004). A seminal paper by Zhou *et al.* utilized UBM to acquire cardiac data in the normal murine heart (Zhou *et al.*, 2004). The same study established three acoustic windows and a reference data table for common echocardiographic parameters in mice, with comparisons made between parasternal long-axis and short-axis views (**Table II**).

An example of the ability for echocardiography to assess disease states is found in a study by Drolet *et al.*, where aortic valve stenosis, as diagnosed by 2-D echocardiography in a parasternal long-axis view, was found in wild-type and LDL receptor -/- mice fed a high-fat/high-carbohydrate diet. The mice displayed smaller valve areas and thicker aortic valve leaflets due to lipid and macrophage infiltration after four months on the regimen (Drolet *et al.*, 2006).

Table II. Native Murine Heart Morphology and Dimensions by UBM in Long- and Short-Axis Views*

Parameters	Right Heart	Left Heart	P value
	Left Ventricle (Long-axis view)	*Left Ventricle (Short-axis view)*	
Heart Rate (BPM)	429±19	429±18	
End-Diastole			
AW (mm)	0.86±0.03	0.84±0.03	NS
EDD (mm)	3.84±0.09	3.80±0.07	NS
PW (mm)	0.74±0.02	0.75±0.03	NS
End-Systole			
AW (mm)	1.36±0.07	1.34±0.06	NS
ESD (mm)	2.45±0.09	2.44±0.08	NS
PW (mm)	1.08±0.04	1.07±0.04	NS
PW velocity (mm/s)	18.6±1.0	17.9±1.0	NS
	AAo (left parasternal)	*AAo (right parasternal)*	
Heart Rate (BPM)	404±16	407±20	
ESD (mm)	1.37±0.02	1.36±0.04	NS
EDD (mm)	1.18±0.01	1.4±0.04	NS

*Adapted from *Zhou et al.*, *Physiol. Genomics. 18(2):232-44 (2004)*.

n = 17 (C57BL/6). Values are reported as mean ± SEM. $P<0.05$. *NS*, not significantly different. Comparisons were made using paired *t*-tests.

AW, anterior wall thickness; *EDD*, left ventricular end-diastolic daimeter; *PW*, posterior wall thickness; *ESD*, left ventricular end-systolic daimeter

AAo, ascencing aorta

1.3.4.7   Systolic and Diastolic Function

Of the systolic indices of cardiac function employed in murine echocardiography, % fractional shortening (%FS) and % ejection fraction (%EF) are the most common; these measures however are load-dependent and cannot differentiate contractile changes from altered loading conditions (Hoit *et al.*, 1995; Pollick *et al.*, 1995). The derived maximum rate of pressure generation (dP/dtmax) has also been commonly used, though it is susceptible to the same aforementioned problem. Load-independent indices of LV performance, such as the end-systolic wall stress (σ_{es}) rate-corrected velocity of fiber shortening (Vcf$_c$) relationship, have been employed in mouse echocardiography (Fentzke *et al.*, 2001), but require extensive mathematical manipulations and invasive procedures to determine aortic pressure. More commonly, cardiac output (CO) has been used to evaluate systolic function and may be determined as the difference between 2D-derived LV end-systolic and end-diastolic volumes (Janssen *et al.*, 2002). A CO-derived parameter, stroke volume (SV), has also been assessed using various acoustic windows and approaches, though limitations in conventional systems, such as poor beam-to-flow alignment and difficulty in determining aortic diameter accurately, have forced the use of CO, SV and associated parameters as comparative, serial measurements rather than as absolute figures (Collins *et al.*, 2003).

Recent studies have proposed alternative methods of assessing cardiac systole and diastole, and have included tissue Doppler imaging to determine mitral annular velocities (Schaefer *et al.*, 2003) and strain rate, which measures the velocity gradient of two points over a fixed vessel segment (Sebag *et al.*, 2005); the latter study showed that tissue Doppler indices increased with administration of dobutamine and decreased with administration of esmolol and that both indices correlated strongly with LV change in maximum pressure development in healthy mice, illustrating the potential of tissue Doppler imaging for the sensitive indication of LV global and regional function (Sebag *et al.*, 2005). Morphologic studies investigating LV hypertrophy as a result of valvular disease and aortic constriction have been made possible by Colour Doppler modality in a variety of transgenic mouse models, but are limited by low frame rates (Collins *et al.*, 2003). Pulsed-wave Doppler modality, such as the one employed by the current study's UBM system, provides transvalvular flow velocity waveforms that can conveniently determine aortic and pulmonic peak velocities, isovolumic relaxation and contraction times (IVRT, IVCT), Doppler flow velocities (E/A ratios), LV ejection times, and mitral inflow velocity patterns. Evaluating diastolic function in murine models strictly using Doppler flow velocities (E, A, E/A ratios) is problematic since in order to differentiate between the E and A waves, the heart rate needs to be significantly slowed down, thus inviting criticism to the relevance of such findings (Scherrer-Crosbie, 2006). A combination of figures to calculate the Tei index – a myocardial performance index that combines diastolic and systolic cardiac function (Tei, 1995) – has been purported to be a more reliable measurement (Collins *et al.*, 2003).

Utilization of serial echocardiography to evaluate the effects of disease on systolic and diastolic function has been accomplished in transgenic mice expressing a dominant-negative form of the CREB transcription factor (CREBA133), which develop dilated cardiomyopathy, attenuated LV systolic function and abnormal diastolic relaxation – all of which closely resemble the disease in humans (Fentzke *et al.*, 2001). Recently, UBM imaging employing a high-frequency 30 MHz probe has been used to assess cytoprotective signalling via TNF receptors against drug-induced systolic dysfunction in TNF-receptor deficient mice (Lien *et al.*, 2006).

1.3.4.8   Blood Flow

Conventional echo systems have traditionally experienced difficulty in measuring aortic and mitral flows, which are typically generated from the parasternal long axis view due to the small size of arterial vessels and valvular orifices compared to the probes as well as the misalignment with the direction of flow (Scherrer-Crosbie, 2006). Conversely, UBM systems utilizing transthoracic high-frequency probes and pulsed Doppler modality are exquisitely capable of characterizing cardiac flow patterns and haemodynamics (Zhou *et al.*, 2004). These flow patterns include atrial and ventricular inflow and outflow tracts, left coronary artery (LCA) and the aortic arch and its three branches. **Table III** shows a comparison of arterial and transvalvular Doppler velocities using different imaging orientations.

Recently, investigators have employed UBM-derived haemodynamics with a 40 MHz probe to assess the degree of coronary artery atherosclerosis by calculating velocity ratios of the proximal to middle LCA in apoE/LDLR double knock-out mice (Gronros *et al.*, 2006). UBM can also provide functional information on cardiac vasculature as evidenced by a recent study by Wikstrom *et al.*, in which a 40 MHz probe was used to measure atherosclerosis-related minimal lumen diameter narrowing of the LCA in living LDLR -/- mice (Wikstrom *et al.*, 2005). These studies and others illustrate the vast potential of UBM imaging in elucidating mechanisms of disease, by employing non-invasive, serial, high-resolution imaging of experimental mouse models in cardiovascular research.

Table III. Native Murine Heart Haemodynamics by Pulsed Doppler UBM and Intracardiac Comparison of Flow Velocities*

Parameters	Right Heart	Left Heart	P value
	Right Superior Vena Cava	*Pulmonary Vein*	
Heart Rate (BPM)	381±11	441±16	
Doppler Waveform			
Peak S (cm/s)	19±2	20±1	NS
Peak D (cm/s)	12±1	54±4	<0.01
Peak A (cm/s)	6±1	4±1	NS
	Tricuspid Orifice	*Mitral Orifice*	
Heart Rate (BPM)	389±22	405±18	
Doppler Waveform			
Peak E (cm/s)	28±1	60±2	<0.01
Peak A (cm/s)	38±1	39±2	NS
Peak E/A ratio	0.73±0.04	1.58±0.07	<0.01
	Main Pulmonary Artery	*Ascending Aorta*	
Heart Rate (BPM)	435±16	381±16	
Doppler Waveform			
Peak Flow Velocity (cm/s)	66±2	105±2	<0.01
Time-velocity Integral (cm)	2.9±0.1	4.4±0.1	<0.01

*Adapted from Zhou et al., Physiol Genomics. 18(2):232-44 (2004).

n = 17 (C57BL/6). Values are reported as mean ± SEM. P<0.05. NS, not significantly different. Comparisons were made using paired t-tests. Peak S, velocity of systolic wave (S-wave) of the superior vena caval or pulmonary vein flow spectra. Peak D, peak velocity of the early diastolic wave (D-wave) of superior vena caval or pulmonary vein flow spectra. Peak A, peak velocity of late diastolic retrograde wave (A-wave) due to atrial contraction in vena caval and pulmonary vein flow spectra OR peak velocity of late diastolic filling wave of mitral/tricuspid Doppler flow spectra; Peak E, peak velocity of early diastolic filling wave (E-wave) of mitral/tricuspid Doppler flow spectra. Peak E/A, ratio of peak E to peak A of mitral/tricuspid Doppler flow spectrum.

1.3.5 Echocardiography of Heterotopic Cardiac Graft Rejection

1.3.5.1 Diastolic Dysfunction

Changes in diastolic measurements have been reported to precede those of systolic indices of graft status in clinical rejection (Desruennes *et al.*, 1988; Amende *et al.*, 1990; Furniss *et al.*, 1991). However, previous studies on murine HHTx grafts indicate that both the left and right atria do not participate in atrial filling and thus are not subsequently involved in ventricular filling during diastole (Zhou *et al.*, 2007; Scherrer-Crosbie *et al.*, 2002). The recommended indices of diastolic performance (*e.g.*, peak early mitral and tricuspid diastolic flow velocity, mitral and tricuspid E/A ratios, mitral A-wave duration) are therefore irrelevant to this model and/or expected not to yield scientifically important information regarding graft status as it pertains to rejection. However, isovolumic relaxation time (IVRT), which was shown to be independently associated with clinical moderate to severe graft rejection (Furniss *et al.*, 1991), has been tested using a similar parameter in rat

HHTx grafts - the relaxation time constant (T_E) of the LV pressure fall - to evaluate left ventricular diastolic dysfunction during allograft rejection (Szabo *et al.*, 2001). Both studies reported a prolonged relaxation time in association with histological rejection when compared to control isografts. Using implanted balloons, the latter study demonstrated that significant decreases in T_E can be detected as early as three days PTx in allografts versus isografts. The authors suggest that this finding was indicative of impaired myocardial relaxation and was in association with decreased myocardial blood flow and increased LV end-diastolic pressure. Other clinical as well as experimental studies have also noted the usefulness of evaluating the maximum rate of LV pressure rise (dP/$t$max), which has been shown to be associated with higher incidences of rejection in humans (Valantine *et al.*, 1989) though the data in rat HHTx allografts has yielded conflicting results (Yoshida *et al.*, 1998; Dambrin *et al.*, 1999; Yoshida *et al.*, 1998). Moreover, reports on the validity of assessing peak systolic and diastolic LV pressures in rat HHTx grafts indicate that both quantities decrease significantly in allografts on day 5 PTx, whereas they remain relatively unchanged in isografts (Szabo *et al.*, 2001). However, these measurements require invasive procedures, such as implanting a balloon in the cardiac graft to measure pressure-volume relations. Finally, reduced left ventricular mean filling volume was demonstrated in day 5 rat HHTx allografts. The authors of the study suggested that abnormalities of diastolic properties previously attributed to the unloaded state of non-working heart models may actually reflect inadequate peri-transplantation myocardial protection (Soto *et al.*, 1998).

1.3.5.2   *Systolic Dysfunction*

Left ventricular parameters of systolic function have proved to be of limited use in assessing graft rejection in the clinical arena (Vivekananthan *et al.*, 2002), and the general consensus is that they are inferior to the diagnostic potential of diastolic changes (Burgess *et al.*, 2002) and are not rejection-specific (Sade *et al.*, 2006). However, significant decreases in % ejection fraction and stroke volume have been associated with histologic rejection (Sade *et al.*, 2006). In rat HHTx allografts, diminished cardiac output and stroke work after isoproterenol treatment was found on day 3 PTx. (DiSesa *et al.*, 1991). Furthermore, studies in HHTx murine grafts indicate a diminishing trend in %EF and %FS in allografts from days 3 to 5 PTx (Scherrer-Crosbie *et al.*, 2002). In our recent study employing UBM-Doppler by

41

Zhou *et al.,* left ventricular %FS was found to increase significantly in the long-term, from day 14 to 50 in isografts, though no parallel study was conducted in allografts (Zhou *et al.,* 2007).

1.3.5.3 Coronary Perfusion

Severe impairment of coronary reserve assessed by thermodilution catheterization has been associated with rejection in patients with orthotopic heart transplants and detection of left ventricular dysfunction has been associated with transplant coronary artery disease (Dandel *et al.,* 2003; Nitenberg *et al.,* 1989). In particular, coronary endothelial function has also been shown to predict progression of cardiac allograft vasculopathy after heart transplantation (Hollenberg *et al.,* 2004). Though cardiac allograft vasculopathy is a long-term complication and does not reflect the pathological changes seen in the short time span of non-immunosuppressed acutely rejecting grafts, from a physiologic viewpoint, graft perfusion is essential for graft viability (Kevin Wei, 2004). Studies in rat HHTx models have shown a marked decreased in myocardial tissue blood flow as assessed by the hydrogen-clearance method in day 3 and 5 allografts, whereas it remained stable in isografts (Szabo *et al.,* 2001). Using UBM-Doppler and M-mode analyses, our group previously showed that after 7 days PTx in murine HHTx allografts, forward and backward blood flow in the LCA of allografts is significantly diminished (Zhou *et al.,* 2007). This is in stark contrast to what is seen in isografts, where the maximal velocity of forward flow to perfuse the myocardium significantly increased from day 1 to 5 and remained consistent after 2 weeks post-implantation (Zhou *et al.,* 2007). The same study also characterized a previously undocumented Doppler waveform of the LCA in murine HHTx grafts, showing the complex variations and periodic patterns as a result of host circulatory pressure and graft contractions (Zhou *et al.,* 2007).

1.3.5.4 Valvular Dysfunction

Tricuspid regurgitation, which has been shown to be prevalent in orthotopic grafts due to elevated pulmonary vascular resistance and secondary ventricular dysfunction, is associated with poor long-term graft survival (Anderson *et al.,* 2004). The events responsible for inducing tricuspid regurgitation are not seemingly applicable in the mouse model. The right atrium and ventricle, very much like the left side of the heart, are under conditions of low pressure due to non-loading conditions, which results in myocardial atrophy (Mottram *et al.,* 1988), inevitably resulting in right atrial and ventricular collapse. Furthermore, we previously showed in murine HHTx isografts that tricuspid and pulmonary waveforms are small in amplitude and variable in waveform obtained by UBM-Doppler, thus precluding these measurements as reliable parameters of evaluating rejection (Zhou *et al.,* 2007). The aforementioned collaborative study also characterized mitral and aortic regurgitation in isograft control studies, which showed high regurgitant jets PTx, with consistently high values (>200 cm/s) achieved in isografts after two weeks post-implantation (Zhou *et al.,* 2007).

1.3.5.5   *M-mode and Graft Morphology*

Interstitial edema and secondary increases in myocardial mass and wall thicknesses have seemingly become obsolete manifestations of clinical rejection due to the effects of immunosuppressants in decreasing edema and immune cell infiltration (Burgess *et al.,* 2002). Furthermore, early morphological changes in the graft are usually associated with the effects of ischemia. However, late acute increases in left ventricular mass and myocardial thickness may be correlated with advanced cellular rejection (Asante-Korang, 2004; Del Castillo *et al.,* 1989), and increases in graft mass in rejecting cardiac grafts were a well-known phenomenon in the pre-cyclosporine era (Burgess *et al.,* 2002). Decreases in LV dimensions were also observed to occur significantly more in rejecting than in non-rejecting grafts (Dodd *et al.,* 1993). Similarly, in a previous study of mouse HHTx grafts that did not utilize immunosuppressive agents, serial echocardiography detected a significant increase in LV posterior wall thickness and conversely, a decrease in LV end-diastolic diameter in rejecting allografts from days 0-5 PTx versus isografts (Scherrer-Crosbie *et al.,* 2002). Unfortunately, the poor resolution at lower frequencies and the high inter- and intra-observer variability have limited interpretation and applicability of these findings. Furthermore, this observation should be taken cautiously, since ventricular atrophy due to the non-loading nature of the graft can result in decreased myocardial wall thicknesses in isografts (Welsh *et al.,* 2001a; Spencer *et al.,* 2003). Finally, studies demonstrate that interventricular septal thickness in cervically-placed heterotopic canine cardiac grafts increased significantly on day 5 PTx

43

(Ducharme *et al.,* 1999). Utilizing UBM-derived M-mode imaging, our team found that in isografts, LV anterior wall thicknesses were consistent throughout the post-transplantation period, but posterior wall thicknesses gradually decreased and the LV chamber dimensions were reduced (Zhou *et al.,* 2007).

Table IV summarizes the findings on the various imaging parameters established by echocardiography in evaluating graft rejection and the proposed mechanisms suggested by the authors of the respective study for each result, as well as the documented or anticipated directionality of such parameters in the murine HHTx model.

Table IV. Summary of Experimental Echocardiography in Heterotopic Animal Models and Applicability to the Mouse Model

Echo parameters used to evaluate heterotopic (abdominal) acute graft rejection	Directionality and mechanism of change in heterotopic acute graft rejection	Applicability and anticipated changes in Tx mouse model during acute rejection
Heart mass (HETERO-RAT) — J Thorac Cardiovasc Surg. 1994 Nov;108(5):925-37; Am J Physiol Heart Circ Physiol 284 H2061-H2068, 2003	Increased wet/dry weights in allografts on day 5 post-Tx vs. isografts due to ischaemic injury and PMN infiltration. 38% decrease in isograft mass @ day 7	Increased graft mass in allografts due to edema and inflammatory responses and decreased isograft mass due to chronic myocardial atrophy
Left ventricular mass — NOT FOUND IN LITERATURE	NOT FOUND IN LITERATURE	From clinical data, an increase in LV mass occurs in rejecting and non-rejecting grafts early post-Tx, but degree of increase in allografts may be more esp. in severe reject
Aortic regurgitation (HETERO-MOUSE) — Ultrasound Med Biol. 2007 Jun;33(6):870-9.	Found to be consistently elevated in isografts; No parallel study on allografts	UNKNOWN
Tricuspid regurgitation - NOT FOUND IN LITERATURE	NOT FOUND IN LITERATURE	UNKNOWN
Mitral regurgitation (HETERO-MOUSE) — Ultrasound Med Biol. 2007 Jun;33(6):870-9	Found to be consistently elevated in isografts; No parallel study on allografts	UNKNOWN
%Ejection fraction, % Fractional shortening (HETERO-MOUSE) — Am Soc Echocardiogr. 2002;15:1315-20	%EF and %FS decreased significantly during acute rejection in allografts but differences vs. isografts may not be significant except at severe rejection	Variable due to unfounded nature of graft; significant decreases in %EF and %FS expected at severe rejection
Stroke work, cardiac output (HETERO-RAT) — J Thorac Cardiovasc Surg. 1991 Mar;101(3):446-9	Biopsy-proven rejecting day 3 post-Tx allografts had diminished CO and SW after experimental injection when compared to isografts; due to systolic dysfunction	Significant decreases in severely rejected grafts with clear systolic dysfunction
Posterior wall thickness (HETERO-MOUSE) — J Am Soc Echocardiogr. 2002 Oct;15(10 Pt 2):1315-20	Increases in PWTh in isografts early post-Tx due to IR injury, gradual increase in PWTh post-Tx are associated with histological evidence of myocyte necrosis, cell infiltration	Increases in PWTh is iso- and allografts early post-Tx but consistently increasing PWTh in rejecting allografts
Heart rate variability (HETERO-RAT) — J Heart Lung Transplant 1999;18:488-508	Total power of HR variability in allografts decreased significantly from 1.5 to 6 days post-Tx. Allograft peak-to-peak amplitude of QRS complex and HR were significantly decreased on day 5.5 or later	Decreased QRS amplitude in severely rejecting allografts; power of HR variability may increase during onset of cellular rejection
Relaxation time (HETERO-RAT) — World J Surg 2001; 25(5) 545-552	Prolonged relaxation time in histologically rejecting grafts starting at 3 days post-Tx, may be related to histological rejection and/or decreased MBF and possibly ischaemia	Histologically rejecting grafts may illustrate impaired myocardial relaxation, i.e. prolonged relaxation time
Atrial reversal velocity - N/A	N/A	N/A
LV peak systolic, end-diastolic pressure (HETERO-RAT) — World J Surg 2001; 25(5) 545-552	Significant decrease in LVPSP at day 5 post-Tx in allografts, stable in isografts. LVEDP increased significantly on day 5 post-Tx in allografts, stable in isografts, etiology not discussed	Decreases in LVPSP may be expected with the onset of obvious systolic dysfunction in allografts; LVEDP may also increase in allografts to indicate impaired compliance
Mitral valve half-pressure time - N/A	N/A	N/A
Mitral deceleration time - N/A	N/A	N/A
Tricuspid inflow (HETERO-MOUSE) — Ultrasound Med Biol 2007 Jun;33(6):870-9	Small and variable in waveform and not measured post-Tx in isografts; No parallel study on allografts	UNKNOWN
Tei Index/Index of Myocardial Performance - NOT FOUND IN LITERATURE	NOT FOUND IN LITERATURE	Significant decreases in IMP in severely rejected grafts with clear systolic and diastolic dysfunction
Intraventricular septal thickness (HETERO-DOG) — J Heart Lung Transplant. 1999 Jun;18(6):510-6	Significant increase in IVS thickness prior to sacrifice (5-7 days) in dog allografts; Etiology not discussed	IVS thickness may be expected to occur predominantly in rejecting allografts though the degree, day of onset, and relative changes vs. isografts are not known
Mean LV filling volume (HETERO-RAT) — J Heart Lung Transplant 1998 Jun;17(6):604-16	Decreased LV filling volume in day 5 post-Tx allografts; Edema and ischaemia possible causes	Decreased LV volume with associated decreases in LV chamber size and thrombus formation in LV; more prominent in allografts vs. isografts
Myocardial echogenicity (HETERO-GOAT) — J Heart Transplant. 1987 Jan-Feb;6(1):21-7	All myocardia increased in brightness at day 2 but decreased thereafter caused by transplant injury/ischaemia	Transient increases in echogenicity and subsequent decrease at advanced rejection
Left ventricular end-diastolic diameter (HETERO-MOUSE) — Am Soc Echocardiogr 2002;15:1315-20	MM decreases in VEDD in isografts but significant in allografts close to day 5 allografts when compared to day 1 post-Tx and isografts, not measured past day 5 post-Tx	Anticipated decreases in LV diameter with assoc. LV hypertrophy during the early onset of cellular rejection; late LV dilation may increase diameter in advanced rejection
A-Av interval (A: mitral late flow; Av: outflow) - NOT FOUND IN LITERATURE	NOT FOUND IN LITERATURE	UNKNOWN
Tricuspid E/A ratio - NOT FOUND IN LITERATURE	NOT FOUND IN LITERATURE	UNKNOWN
Early mitral flow velocity (E) - NOT FOUND IN LITERATURE	NOT FOUND IN LITERATURE	UNKNOWN
Pericardial effusion - NOT FOUND IN LITERATURE	NOT FOUND IN LITERATURE	N/A
Mitral A-wave duration - N/A	N/A	N/A
Mitral E/A ratio - NOT FOUND IN LITERATURE	NOT FOUND IN LITERATURE	NOT FOUND IN LITERATURE
Myocardial blood (coronary) flow (HETERO-RAT) — J Heart Transplant 8:46 1989	Significant decrease in myocardial coronary flow on day 3 post-Tx but no changes after day 5 in allografts, stable for isografts; tacrolimic rejection and increased edema assoc. w/ reduced MBF	Decreased coronary flow in severe or late rejection
Pulmonary vein systolic flow - N/A	N/A	N/A
Superior vena caval systolic flow - N/A	N/A	N/A
DP/max values (max rate of LV pressure rise) (HETERO-RAT) — J Heart Lung Transplant. 1999 Jun;18(6):524-31	Significant changes in DP/max values in day 5 allografts vs. day 1 values and isografts, isografts remained stable during period of observation	Similar changes in DP/max values may be expected in the model (decreased values in acutely rejecting allografts)

1.4   HYPOTHESES AND OVERALL OBJECTIVES

1.4.1   Objectives and Rationale

The primary objective of this project was first, to investigate specific parameters – those established by clinical echocardiography and available using a novel high-frequency ultrasound imaging modality – that can discriminate between rejecting allografts and control isografts in the PTx period earlier than the commonly used method of finger palpation.

The second objective was to establish a timeline of changes in blood flow (haemodynamics), dimensions and morphology, and overall physiology in a murine model of heterotopic cardiac transplantation. This would allow for more sensitive and earlier determinations of graft viability during different time points following transplantation and allow for a more accurate determination of the effects of various pharmacological, genetic, or molecular treatments and regimens on the immunopathology of graft rejection.

The overall aim was to identify specific and sensitive indices of cardiac function in heterotopic cardiac grafts that would be exceptionally useful for evaluating the dynamic effects of immunologically-mediated events on graft physiology, particularly in the setting where immunological agents, pharmacological treatments, and genetic modifications that are capable of modulating the immune response to allografts can be tested in a model of experimental heart transplantation. The value of establishing such a method is particularly appreciated with the advent of the genomics and proteomics era, where specific gene knock-out and transgenic models have now become available, and are currently being produced in the mouse model. These discoveries have paved the way for molecular biologists, clinical biochemists, and transplant scientists to elucidate specific gene-protein pathways and the components of the immune system that are implicated in the rejection process (Honjo *et al.,* 2004). This information can lead to further investigations of the effects of pharmacological agents on graft survival (Tanaka *et al.,* 2005b). Taken together, studies on rodent HHTx are exceptionally relevant for understanding the basic science of immunopathology of graft rejection, examining organ/ischemia reperfusion injury and graft preservation, allowing a first-line evaluation for novel pharmacological agents, and paving the way for the development of new technologies for application in the transplantation of vascularized organs (Cramer *et al.,* 1994).

46

1.4.2 Characteristics of Target Imaging Parameters

The ultrasound measurement or group of measurements in question should ideally possess the following eight attributes:

(1) Be easily acquired using a high-frequency imaging unit appropriate for small animal imaging;

(2) Be limited in the amount of off-line mathematical or computer manipulations required;

(3) Be acquired relatively quickly so as to allow for the evaluation of a large number of specimens for study;

(4) be able to discriminate statistically significant differences between rejecting and non-rejecting heterotopic grafts early PTx, before large changes in finger palpation scores of the graft or substantial histological changes can be seen;

(5) Have sensitivity and specificity for detecting rejection greater than, or at least equal to, the current methods of palpation and histology;

(6) Be obtained non-invasively which would allow serial or specific time-point evaluations to be made in the PTx period without adversely affecting the test subject;

(7) Not be susceptible to large fluctuations in the post-ischaemic period and be consistently indicative of graft status early, mid- and late in the progression of rejection; and

(8) Is physiologically relevant such that the measured change(s) in the parameters is/are at least partly explained by immunological mechanisms.

1.4.3 Hypotheses and Specific Aims

The hypotheses and specific aims put forth in this study are three-fold:

Hypothesis #1: *UBM can provide a comprehensive, high-resolution assessment of the rejection process in heterotopic cardiac murine grafts.*

Specific Aim #1: To characterize previously undocumented changes in haemodynamics, cardiac morphology, and overall graft physiology in a commonly used model of murine heterotopic cardiac graft transplantation during acute cardiac rejection using serial, high-resolution UBM imaging and derived functions. These include, but not limited to, the effects of ischemia-reperfusion injury, cardiac unloading, and the interaction of native and graft cardiac cycles.

Hypothesis #2: *Echocardiographic changes assessed by UBM imaging can delineate temporal changes in accepting and rejecting murine cardiac grafts.*

Specific Aim #2: To establish a timeline of the dynamic changes in haemodynamics, cardiac morphology and graft physiology that take place during acute cardiac graft rejection in the above-named model using UBM. This endeavour would put into context past and future work in transplantation research where it is acknowledged that cardiac physiology, morphology and blood flow are important determinants of graft viability and that change over time during graft rejection;

Hypothesis #3: *UBM-derived parameters can non-invasively detect cardiac graft rejection.*

Specific Aim #3: To elucidate which haemodynamic, morphological, and/or physiologic parameters assessed with a novel UBM imaging system can differentiate between control isografts and acutely rejecting allografts in the above-mentioned model. Utilizing echocardiographic parameters established by human echocardiography and modified by our laboratories for this model system, serial UBM imaging of syngeneic and allogeneic transplants over the post-transplant course would help delineate particular changes allografts and isografts undergo in response to immunological rejection (or lack thereof) and potentially give rise to sensitive and specific UBM-derived predictors of rejection in this model.

Given the current state of the literature and our previous work on isografts (Zhou *et al.,* 2007), it is hypothesized that the most reliable echocardiographic parameters that can potentially discriminate between rejecting and non-rejecting grafts are:

a) Decreased Doppler velocities in coronary arteries due to the anticipated decrease in graft perfusion in acutely rejecting grafts;

b) Diminished systolic function in acutely rejecting grafts as reflected by decreases in ejection fraction (%EF), %fractional shortening (%FS), stroke volume (SV) and cardiac output (CO);

c) Increased graft mass due to interstitial edema and immune cell tissue infiltration that are expected to occur during allograft rejection;

48

d) Increased wall thicknesses of the left ventricle due to inflammatory processes and myocardial edema, compensatory hypertrophy, and secondary changes in ventricular dimensions;

e) Decreased graft heart rate as determined by UBM-Doppler and ECG, reflecting pathological changes in allograft myocardial electrophysiological properties; and

f) Increased echocardiographic backscatter due to cellular changes within the graft as a result of histopathological mechanisms (*e.g.,* apoptosis) that may occur during allograft rejection.

II. MATERIALS AND METHODS

2.1 ***Animal Preparation.*** C3H/He (H-2^K) and Balb/c (H-2^d) mice were obtained from Charles River Laboratories (St. Consant, QC, Canada). All housing, treatment protocols, and procedures were reviewed and approved by the Hospital for Sick Children in Toronto Animal Care Committee in accordance with the current regulations and standards established by the Canadian Council of Animal Care.

2.2 ***Heterotopic Cardiac Transplantation.*** Highly trained microsurgeons performed syngeneic (C3H/He->C3H/He, n=9) and allogeneic (Balb/c->C3H/He, n=13) transplants in 6 to 8-week old female C3H recipients using the procedure originally reported by Corry *et al.* (Corry *et al.,* 1973b) and modified by our laboratory (West and Tao, 2002). The donor mice were of similar age and sex-matched, and had similar body weights as the recipients. In short, mice were anaesthetized with 3.6% chloral hydrate (Wiler, London, ON, Canada). The donor aortic root was anastomosed end-to-side to the recipient abdominal aorta and the donor pulmonary artery trunk to the inferior vena cava (IVC). The mice were allowed to recuperate for 24 hours from surgery prior to imaging.

One mouse with an implanted isograft died for reasons unrelated to transplant surgery six days post-transplant. Two isografts and three allografts were excluded from analysis; of those five, a dislodged ventricular thrombus found traversing the graft aortic orifice during cardiac graft systole was observed in two isografts and two allografts and aortic and coronary flow measurements in these grafts were not obtainable; the third allografted mouse sustained a chronic skin infection at the site of incision and resulted in scarring that prevented clear image acquisition. This graft also survived past 40 days, presenting an abnormal pattern of survival in this fully-mismatched donor-recipient murine strain combination.

2.3 ***Ultrasound Biomicroscopy***

2.3.1 *Instrumental specifications.* An ultrasound biomicroscopy (UBM) system (Vevo 770, VisualSonics Inc., Toronto, ON, Canada) was used. It has a single element mechanical transducer with a centre frequency of 30 MHz and a frame rate of 30Hz. The spatial resolution of B-mode imaging was ~115 μm (lateral) by ~55 μm (axial). Focal depth was set at 12.7mm and power at 100%. In M-mode, the repetition rate is 1,000 Hz, with 512 depth samples for an 8-mm sampling window. The Doppler pulse repetition frequency is up to 96 kHz, corresponding to a maximum unaliased velocity of 120 cm/s (with incidence angle being zero). In this study, the axial dimension of the Doppler sample volume was 0.5 mm.

2.3.2 Experimental Plan

Grafts were imaged daily at approximately the same time each day in the afternoon; isografts were imaged for 2 weeks and allografts were imaged until rejection (median survival time = 12 days) as assessed by ECG and echocardiography (see below for criteria for determining graft rejection). We evaluated the physiologic, haemodynamic and morphologic changes in syngeneic and allogeneic transplants over the post-transplant course serially using a battery of echocardiographic parameters used in human echocardiography that were modified by our laboratory for use in this model system.

2.3.3 Animal Preparation.

The mice were fasted for 2-3 hours prior to imaging to reduce gastrointestinal activity and gas from interfering with image acquisition. Only water was supplied during the fasting period. Mice were weighed daily prior to imaging. Anaesthesia was induced by placing the mice in a closed chamber and filled with 5% isoflurane gas in oxygen until the mouse was unconscious (approx. 1 minute). Anaesthesia was maintained thereafter at 1.5% isoflurane and oxygen for the entire duration of the imaging session through a face mask. Anaesthetized mice were laid in supine with all four paws taped to a platform which consisted of commercially built-in electrodes for recording ECG (**Fig. 2**). ECG spectra included cardiac voltage complexes from both native and graft hearts. Abdominal hair was removed using hair-removal cream (Nair, Mississauga, ON, Canada). Pre-warmed ultrasound gel (Aquasonic 100, Parker Laboratories, Inc., Fairfield, NJ, USA) was placed on the abdomen for coupling the transducer and the abdomen. Mouse body temperature was acquired by a rectal thermal probe (Indus Instruments, Houston, TX, USA) feeding to a digital monitor and maintained within the range of 36~38 °C using a heating lamp.

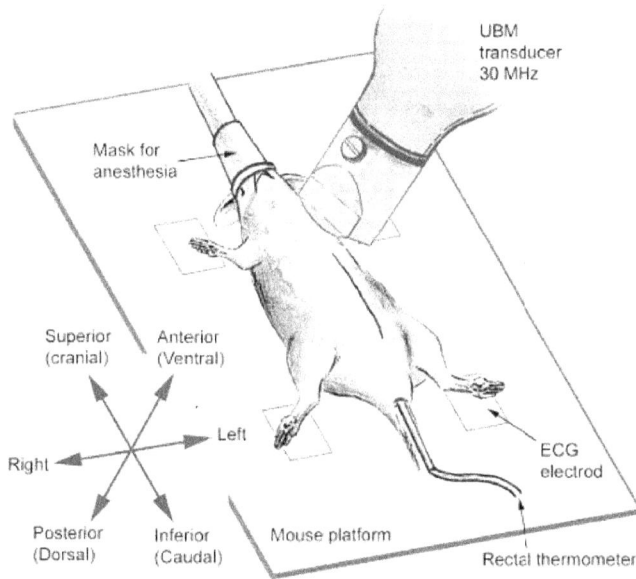

Figure 2 – Experimental Setup. The orientation of the mouse and the position of the UBM transducer are shown. The terms superior, inferior, anterior, and posterior sides correspond to the cranial, caudal, ventral, and dorsal regions of the mouse body, respectively. The surgical line of incision of the abdomen is also drawn to indicate the general location of the graft heart. Adapted by permission from Zhou *et al., Physiol. Genomics. 18(2):232-44 (2004).*

2.3.4 Graft Survival. Graft viability was determined using two methods for comparison: abdominal palpation and UBM-derived parameters. The palpation method was accomplished by daily finger palpations of heart pulsation in unanaesthetized mice (**Fig. 3**). A single blinded observer (R.B.) trained in the finger palpation technique assigned a qualitative score from 4 (vigorous pulsation) to 0 (absence of pulsation) daily to evaluate graft function according to previously reported criteria (Corry *et al.,* 1973b). A value of '0' indicated graft death. The second method utilized echocardiographic determination of graft death and included the following criteria: absence of Doppler waveforms established to be caused by graft systole from the ascending aorta and left coronary artery (LCA) (Zhou *et al.,* 2007), no traceable graft ECG, and visual B-mode confirmation of cessation of ventricular contraction.

52

Generally, all three criteria had to be satisfied in order to pronounce graft death (**Fig. 4**).

2.3.5 Graft Orientation and ECG. Graft orientation was serially determined by finger palpation as day-to-day movements can often change the orientation of the graft within the abdomen. Typically, however, the cardiac graft was found to be positioned to either the right- or left-middle portion of the abdomen, with the cardiac base proximal to the medial or surgical mid-line, and the apex pointing inferiorly and to the lateral right or left direction.

ECG recordings were visualized on the UBM system monitor for all mice using electrodes built into the imaging platform. Abdominal-intracardiac ECG was obtained simultaneously (**Fig. 5A**) and limb ECG (**Fig. 5B**) was also performed using preliminary isografts (n=2) and allografts (n=2) (Gould Instrument System, Ponemah Physiology Platform, Valley View, OH, USA). Abdominal and limb/paw ECG was obtained by needle electrodes, and intracardiac ECG was acquired by implanting a custom biopolar lead into the jugular vein as an access point for advancing the lead through the superior vena cava and into the native right ventricle. This experiment was performed to determine if graft and native heart waveform morphology and rate could be differentiated using the UBM system for accurate acquisition and corroboration of graft HR with Doppler-derived graft HR (**Fig. 6**).

ECG-based kilo-hertz visualization (EKV), a relatively new mode developed for the UBM system, was performed in a small number of preliminary grafts to gauge the applicability of this system in yielding additional information on graft viability and physiology. EKV produces a video cine loop of the cardiac cycle.

2.3.6 Graft Mass. The grafts were weighed on a digital scale (AG104, Mettler Toledo, Mississauga, ON, Canada) before implantation and after explantation to determine myocardial weight changes in isografts and allografts (**Fig. 7**).

2.3.7 Semi-Quantitative Evaluation of Changes in Echogenicity. LAV B-mode images of the LV were used to approximate changes in visual intensity/backscatter seen in the post-transplant period (**Fig. 8**). A segment of the mid portion of the LV anterior wall was chosen and aligned with the focal zone to score semi-quantitatively the grafts on a scale from 1 (low intensity), 2 (medium intensity) or 3 (high intensity) in a blinded fashion. The anterior wall of the LV was chosen since it is nearest the transducer and is less susceptible to attenuation from other cardiac structures. Every effort was made to reproduce consistently the same graft orientation and to ensure that the anterior wall aligned horizontally and within the focal zone. Visual (gain) settings were maintained the same for all grafts.

2.3.8 M-mode: Graft Dimensions and Physiologic Function. Wall thicknesses and ventricular chamber diameters were calculated from M-mode and averaged from both the LAV and SAV (**Figs. 9-14**). A longitudinal-axis view of the cardiac graft was generally the first orientation obtained and allowed for the visualization of blood flow from the anastomosis of the ascending donor aorta with the native abdominal aorta, retrograde flow through the ascending aorta due to native heart systolic pressure, and transvalvular flow across the aortic orifice into the LV outflow tract (**Fig. 10**). For LAV measurements, the LV was oriented directly perpendicular to the direction of the ultrasound beam in B-mode (**Fig. 11A, B**) and the M-mode cursor line was placed at the segment corresponding to the largest LV chamber dimension. Physiologic data was also acquired from the M-mode (see below). Subsequent to a 90° rotation, the transducer was swept towards the apex until the largest LV dimensions were obtained; the same procedure was carried out for obtaining wall thicknesses and physiologic parameters from the SAV (**Fig. 11C, D**). A novel anatomical M-mode function available on the UBM was also tested for both LAV and SAV for a small number of grafts. Anatomical M-mode allows for the acquisition of time course M-mode without the requirement for a perfectly perpendicular angle between the cursor and the tissue interface, which was advantageous for this animal model given the subtle abdominal movements of the cardiac grafts *in vivo* (see **Appendix**).

2.3.9 Doppler Flow. Doppler recordings were obtained from minimum intercept angles; every effort was made to ensure that the ultrasound beam was nearly parallel to the direction of the flow channel. Generally, intercept angles were kept between 0-60°. ECG was regularly recorded for corroboration of Doppler-derived with ECG waveforms and for determining graft survival and graft HR. A complete examination for one mouse (*i.e.,* between anesthesia and arousal) for the trained sonographer can last approximately 20 minutes, with more average times near 30 minutes.

2.3.9.1 Aortic Regurgitation (AR) and Ascending Aortic (AA) Flow: The Doppler sample volume was placed at the aortic orifice and proximal to the side of the LV to obtain a Doppler spectrum of the aortic regurgitant jet (**Figs. 15, 16**). The Doppler sample volume cursor was subsequently positioned at the mid-region of the donor aortic arch to enable recording of the flow velocity spectrum (**Figs. 19-22**).

2.3.9.2 Left and Right Coronary Arteries (LCA, RCA): The longitudinal imaging plane was slightly adjusted to visualize the proximal segment of the LCA and distal to the aortic sinus (**Figs. 23-26**). Sometimes the RCA was also visible from this view but more often required some adjustment to acquire optimal visual and Doppler recordings (**Figs. 27-30**). For both LCA and RCA Doppler recordings, the sample volume cursor was placed 0.5-1.0 mm distal to the origin of the artery from the aortic sinus, but prior to branching of the arteries. As a newly developed function for the UBM system, Power-Doppler was tested for quantifying blood flow velocity and direction in the LCA (see **Appendix**).

2.3.9.3 Tricuspid Transvalvular Flow and Mitral Regurgitation (MR): The transducer was re-located at the graft apex with the central axis of the transducer pointing to the superior, posterior direction and pointing laterally to either the left or right, depending on the orientation of the graft. In this view, both the LV inflow and outflow tracts were viewed, allowing for the visualization of the aortic and mitral valves. The Doppler sample volume was placed at the mitral orifice, slightly on the atrial side for recording mitral regurgitation during systole (**Figs. 17, 18**). Mitral flow during diastole was difficult to obtain and variable in pattern and was therefore not consistently obtained. A subsequent minor rotation of the transducer counter-clockwise allowed for optimal visualization of the tricuspid orifice and acquisition of tricuspid diastolic inflow Doppler spectra (see **Appendix**).

2.3.9.4 Main Pulmonary Arterial (MPA) Flow: From the long-axis view (LAV) of the LV, a 90° rotation yielded the short-axis view (SAV) of the graft. Sweeping the transducer towards the base, visualization of the RV outflow tract and main pulmonary artery (MPA) was achieved. Doppler pulmonary arterial flow spectrum was recorded at the mid-point between the pulmonary orifice and anastomosis of the MPA with the recipient IVC (see **Appendix**).

2.4   *Offline Analysis*

2.4.1   *General*

B-mode, M-mode and Doppler measurements were made using parameters established by clinical echocardiography (Quinones *et al.*, 2002; Schiller *et al.*, 1989; Sahn *et al.*, 1978) for assessing haemodynamics, cardiac morphology and general graft physiology. A pilot study quantifying Doppler-derived parameters for aortic regurgitation showed that there was no significant difference for Doppler data obtained over 3 cardiac cycles in comparison to 1 cardiac cycle (data not shown). Therefore, all Doppler-derived measurements were obtained from 1 cardiac cycle. M-mode recordings of the LV were analyzed for wall thicknesses, chamber dimensions and general physiologic function. M-mode data was taken for 3 cardiac cycles and averaged from both the LAV and SAV. Semi-automated endocardial wall traces of the LV in M-mode allowed for the automatic generation of common physiologic parameters (*i.e.*, %FS, %EF, CO, SV) using advanced cardiovascular analysis software developed specifically for this UBM system.

2.4.2   *Doppler Flow Characterization and Analysis*

Due to the differing heart rates of the native and transplanted hearts, complex flow velocity waveforms with periodically repeated patterns were observed in the AA (**Fig. 19**), LCA (**Fig. 23, Table VI**), and RCA (**Fig. 27**). To resolve the various components of the Doppler spectra, a typical LCA waveform was analyzed using mathematical computation by Y.Q. Zhou and A. Feintuch (Mouse Imaging Centre) (MATLAB, The MathWorks, Inc., Natick, MA, USA). Assuming the Doppler spectra for the AA, LCA and RCA are actually the sum of two periodic waveforms with different frequencies corresponding to the beating rates of the native and transplanted hearts, the velocity waveform for each heart was represented by a sum of sinusoids with arbitrary amplitude and phase of the corresponding frequency and its first, second and third harmonics. Doppler amplitudes and phases were then extracted by numerically fitting the sum of those two waveforms into the actual Doppler

56

spectrum. The resulting two periodic waveforms represent the flow caused by the native circulation (**Fig. 23C**) and by the graft (**Fig. 23D**).

The constituent segments of the Doppler spectrum associated with the cardiac cycle are summarized in **Table VI.** Based on this analysis, the positive (upward) waveforms of highest amplitude appeared during native heart systole concomitant with graft diastole, and were identified as *forward Vmax* (**Fig. 23D**, *a*). Conversely, the negative (downward) waveforms of highest amplitude occurred during native heart diastole and simultaneous graft systole, and are identified as *backward Vmax* (**Fig. 23D**, *d*). The waveforms attributed with producing backward Vmax are therefore caused by graft contraction. The positive waveforms with 'v-shaped notches' represented the transition from end-diastole to early-systole in the recipient heart. The peaks of moderate amplitude following the notches occurred when graft flow approximates zero and represented the real native circulatory peak systolic velocity (**Fig. 23D**, *b*). However, this peak was exempt from further analysis as it was not readily identifiable. The lowest points of the notches represented native heart end-diastolic velocity (*forward Ved*) since it occurred during zero graft flow velocity (**Fig. 23D**, *c*). Finally, no significant physiologic definition has been assigned when both hearts are at end-diastole (**Fig. 23D**, *e*).

2.4.3   Heart Rate (HR) and Resistance Index (RI)

Native and graft HR measurements (**Fig. 6**) were obtained from aortic Doppler blood flow velocity spectra. Native HR was determined from three adjacent maximal forward/positive waveforms whereas graft HR was measured from three adjacent maximal backward/negative waveforms (**Fig. 19B**). In the event that an accurate graft HR reading was not possible due to advanced rejection or inconsistent ECG, supplemented data from LCA spectra was used, though this contributed to less than 2% of all the HR data. The time-velocity integrals (TVI) were measured for the highest forward and backward Doppler flow waveforms. A modified forward Resistance index (RI) was defined because of the unique interaction of the two hearts and was calculated as the difference between *forward Vmax* and *forward Ved* as the percentage of *forward Vmax*.

2.5   Histology

Histology was used to confirm pathology in acutely rejected allografts and was assessed for evidence of cellular infiltrates and myocyte destruction (**Fig. 32**). Histological slides were prepared by the Division of Pathology at The Hospital for Sick Children (Toronto). Isografts (at day 14) and allografts (upon rejection) were explanted, weighed, and fixed in 4% paraformaldehyde for H&E staining. Grafts were paraffin-embedded and cut into halves; one half was further cut into short-axis sections which were later stained with hematoxylin and eosin (H&E); the second half was reserved for TUNEL staining (see below). Histology slides of isografts (n=7) and allografts (n=10) were digitized using an SMZ800 Zoom Stereomicroscope, captured using a Digital Eclipse DXM1200 High-Resolution digital camera, and digitized with an automatic camera tamer ACT-1 software (Nikon, New York, USA).

2.6   TUNEL Assays

Separate graft specimens from those used for histology were then submitted to Division of Pathology at The Hospital for Sick Children (Toronto) for assessment of apoptosis (**Fig. 32**). DNA damaged cells were detected using the *in situ* end labeling assay for TUNEL (terminal deoxynucleotidyl transferase-mediated DNA nick-end labeling) adapted to an automated immuno/in situ hybridization technique (Discovery Ventana Medical Systems, Inc., Tucson, AZ). Before staining, the paraffin-embedded blocks were sliced into 5 micron-thick sections and blocked for endogenous peroxidase and with a biotin block as well as a subsequent online digestion with protease I (Ventana Medical Systems) for 12 minutes. The assay uses recombinant terminal deoxynucleotidyl transferase (Tdt) (Invitrogen, Carlsbad, CA) for adding homo-polymer tails to the 3' ends of cleaved DNA, characteristic in cells undergoing programmed cell death. Biotin 16-dUTP (Roche Diagnostics, Basel, Switzerland) was the label used for this reaction. Colorimetric visualization using avidin-horse radish peroxidase and 3,3'-diaminobenzidine (DAB) detection method was performed. The counterstain was hematoxylin. Controls included recipient heart (negative) and human thymus (positive) specimens. TUNEL slides were digitized as before (see above).

2.7 Statistics

All statistical analyses and plots were carried out using SigmaStat (SigmaStat, Systat Software Inc., San Jose, CA, USA). Kaplan-Meier analysis was used to estimate graft survival and median survival time, and the Log-rank test was used to test for survival differences in allografts using palpation or echocardiography. Differences in graft weights between isografts and allografts were determined using Student's *t*-test and within each treatment group by the paired *t*-test. Two-way analysis of variance with repeated measures (ANOVA-RM) was used to test for significant differences between iso- and allografts in the serial Doppler, M-mode and echogenicity data in the PTx period, and was followed by post hoc testing by the Holm-Sidak method for all pair-wise comparisons. Significant temporal changes within each group of grafts were determined by 1-way ANOVA-RM. To determine inter- and intraobserver variability, a total of 432 measurements were calculated on 18 mice on the same day post-transplant by two observers. One observer (R.B.) repeated 216 measurements in M-mode in both SAV and LAV several months after the initial data acquisition to calculate intraobserver variability. A second observer (Y.Q.Z) repeated those measurements to determine interobserver variability. Interobserver and intraobserver variability were calculated as the absolute value of the difference between the two observations divided by the average of those values. Observer variabilities are expressed as the value of the variability ± standard deviation (Scherrer-Crosbie *et al.*, 1998). All data are expressed as mean ± standard error of the mean (SEM) and statistical significance was set at $p < 0.05$.

SECTION III. RESULTS

3.1 Observer Variability

Because of changes in the image quality and backscatter over the post-transplant period, determining epicardial and endocardial borders generally became more difficult as time progressed. The interobserver variability was determined to be 0.129 ± 0.284 mm and the intraobserver variability 0.06 ± 0.151 mm from day 2 post-transplant (PTx) M-mode measurements of the left ventricle.

3.2 Palpation

Figure 3 shows mean palpation scores in isografts and allografts as assessed by finger palpable activity of the intensity of ventricular pulsation. Both isografts (n=7) and allografts (n=10) had a maximal mean palpation score of 4 at day 1 PTx, which indicated strong ventricular pulsation. A gradual and parallel decrease from day 1 to day 5 PTx was observed in both groups. The mean palpation score of allografts was significantly less than isografts from day 7 onwards ($P \leq 0.001$).

3.3 Graft Survival

Graft survival was assessed by the palpation method and by echocardiography (**Fig. 4**). Graft survival was different between isografts and allografts by each respective method (palpation, P=0.002; echocardiography, P=0.003) but survival of allografts was not different between methods. **Table V** summarizes the rejection pattern of allografts and the median survival time using palpation or echocardiography. The median survival time by palpation and by echocardiography was 10 and 12 days, respectively, though this was not significantly different.

3.4 Heart Rate (HR) and ECG

Simultaneous abdominal-intracardiac ECG (**Fig. 5A**) and limb ECG (**Fig. 5B**) were measured in mice implanted with isografts (n=2) or allografts (n=2). This experiment was conducted to determine if differences in ECG waveform morphology, rate and amplitude could be potentially verified using the UBM system for accurate corroboration of limb ECG-derived graft HR with Doppler-derived graft HR. Abdominal ECG (**Fig. 5A**, top panel) shows a different rate, amplitude and morphology from the intracardiac ECG spectrum, which is presumably due to the recipient heart (**Fig. 5A,** bottom panel). Limb ECG (**Fig. 5B**) yielded a spectrum seemingly composed of two different waveforms showing distinct amplitudes, morphologies, and rates. One of these waveforms had a similar HR to the abdominal ECG spectrum (indicated as 'Tx', graft) whereas the HR of the other waveform was similar to the intracardiac ECG spectrum (indicated as 'Nv', native heart). Cardiac arrhythmia was detected in allografts late after transplant (**Fig. 5C**, day 10).

Heart rate of the native recipient heart as determined by UBM-Doppler did not differ between isograft and allograft recipients (**Fig. 6A**). Graft HR (**Fig. 6B**) was similar for isografts and allografts from days 1-6, then allograft HR gradually significantly decreased versus isografts from day 7 onwards (*$P{\leq}0.001$). A significant increase in graft HR was seen in both groups of grafts from days 1-3 (Ψ, $P{\leq}0.001$) and day 3 values for allografts were significantly higher than day 14 values (Δ, $P{\leq}0.001$).

3.5.   Myocardial Mass

There was a significant difference in the change in myocardial weight from implantation to explantation between rejected allografts (n=5) versus isografts (n=5) (**Fig. 7**). Weight gain was significantly greater in allografts than in isografts (raw weight gain, $P=0.010$; % weight gain, $P=0.009$). Furthermore, in contrast to isografts, allograft myocardial post-explantation weight increased significantly from pre-implantation values (**$P=0.031$).

3.6   Myocardial Backscatter and Thrombosis

A non-homogeneous increase in echogenicity of the myocardium was observed in both isografts and allografts. **Figure 8** shows a significant increase in semi-qualitative mean scores of myocardial intensity on day 1 versus days 3, 6 and 10 PTx for both groups of grafts (isografts *$P{\leq}0.001$; allografts, **$P{\leq}0.001$), though there were no differences between the two groups in each respective day and no intensity changes after day 1.

Ventricular thrombosis was observed in post-transplant for both isografts and allografts (**Figs. 9** and **10**; compare days 1 vs. 3 and 6). Fresh and hypoechogenic ventricular thrombus was ubiquitous in all grafts in the early post-operative period and occupied most of the LV chamber, allowing only a small volume of blood to flow through the aortic orifice. Generally, at approximately 3 days to 1 week post-transplant, it was observed from serial B-mode images that ventricular thrombi decreased in size, transformed from a pulpous mass to a hardened solid, and became non-homogenously hyperechogenic; this pattern continued late after transplant (>12 days) in isografts, resulting in increased visually identifiable blood flow. In allografts, however, it was observed that LV cavity thrombi ostensibly dissipated late after transplant and replaced by recipient blood (not measured). Because of the presence of ventricular thrombus and increased tissue and thrombus backscatter, M-mode recording and analysis became generally more difficult after day 1, but was still feasible if corroborated with the respective real-time 2-D B-mode image.

3.7   Graft Physiology, Morphology and Dimensions

3.7.1   Cardiac Output (CO), Stroke Volume (SV), %Ejection Fraction (%EF) and %Fractional Shortening (%FS)

M-mode semi-automated endocardial traces of the LV allowed for the generation of echocardiographic parameters of cardiac function (**Fig. 11**). The plots for these load-dependent parameters were highly variable, showed no or little difference between iso- and allografts, and were difficult to interpret because of the large error bars and inconsistent patterns observed post-transplant (see **Appendix**).

3.7.2   Anterior and Posterior Wall Thicknesses

Differences in anterior wall thicknesses at end-systole (AWes) and end-diastole (AWed) were observed between allografts and isografts (*$P \leq 0.001$). Significantly thicker AWes (**Fig. 12A**) and AWed (**Fig. 12B**) thicknesses were observed for allografts from 3 to 10 days and at day 14 PTx in comparison to isografts. Peak allograft AWes and AWed thicknesses were achieved on day 5 and day 7, respectively, but decreased gradually thereafter to values comparable to early post-transplant figures. Generally, isograft anterior wall thicknesses were stable over the study interval (**Fig. 12A, B**). Significant changes in posterior wall thicknesses at end-systole (PWes) and end-diastole (PWed) were also observed between the groups of grafts (**Fig. 13**). Posterior wall thicknesses were significantly higher in allografts versus isografts from day 2 onwards (*$P \leq 0.001$). Allografts reached peak PWes (**13A**) and PWed (**13B**) thicknesses on day 6 PTx, which differed significantly from

62

the lower day 13 or 14 values for both measurements (Δ, $P \leq 0.001$). Allografts showed differences from early (day 1) vs. mid PTx values (days 6, 7) for both parameters (Ψ, $P \leq 0.001$). Isografts also undergo significant changes over time; differences were detected between the higher values for days 3-7 vs. the lower day 13 or 14 figures for both PWes (**13A**) and PWed (**13B**) (Δ, $P \leq 0.001$).

3.7.3 Ventricular Dimensions

End-systolic (**Fig. 14A**, ESD) and end-diastolic (**Fig.14B**, EDD) LV diameter was similar for both isografts and allografts for days 1-9 PTx. Except for the early post-operative phase (days 1-3), ESD and EDD in isografts was generally stable throughout the study period. An early trend towards decreasing diameter was seen in both types of grafts (isografts, days 1-3; allografts, days 1-4) and may be due to the interference of the ventricular thrombus with systolic contraction and diastolic relaxation (see **Fig. 9B, D**). ESD and EDD in allografts increased significantly in comparison to isografts from day 10 onwards. Allograft ventricular dimensions also significantly increased in comparison to mid-transplant values (Δ, $P \leq 0.001$) and reached peak diameter late after transplantation.

3.8 Valvular Regurgitation and Flow

3.8.1 Aortic Regurgitation

Sample 2-D image and Doppler flow spectra of aortic regurgitation (AR) are shown in **Fig. 15**. Periodic gaps in the waveforms were primarily due to respiration. The waveform changes in the AR Doppler spectra are illustrated in **Fig. 16A**. In allografts, AR waveforms are characterized by narrowing Doppler waveform morphology and an irregular pattern of the number of maximal velocity peaks. This was likely a result of aortic insufficiency and widening of the aortic root, which gave rise to a dysfunctional aortic valve as verified by the lack of aortic valvular action seen in 2D imaging. Despite these elements, peak AR in allografts was reached on day 6 PTx and continued to reach <2000 mm/s late after transplant (**Fig. 16B**). Isografts showed no significant differences over time and remained stable during the post-transplant period. Allografts differed from isografts on day 14 only (*P=0.023) though peak AR on day 6 differed significantly from late post-transplant figures (Δ, P=0.001).

63

3.8.2 Mitral Regurgitation

Sample 2-D image and Doppler spectra for mitral regurgitation (MR) was acquired from a sample volume between the LA and LV and just proximal to the side of the LA chamber (**Fig. 17A, B**). Dilatation of the LA was a common, progressive observation seen predominantly in isografts (**Fig.18A, 31A**); this facilitated the acquisition of a MR jet sample volume at the mitral valve. Absence of LA dilatation in allografts, however, made obtaining an MR spectrum difficult and thus several sample volume locations were attempted in such circumstances; if no MR was detected, a value of '0' was assigned. Isografts and allografts had comparable MR velocities ($\sim$1000-2000 mm/s) for 7 days after transplant, until day 8 when there was a significant divergence between the two groups (*$P\leq0.001$). Higher MR velocities were seen in isografts in comparison to allografts after day 7 (**Fig. 18B**). Mitral Doppler inflow (i.e., E-wave A-wave) was not analyzed due to the inconsistency and visual difficulty of acquiring an adequate sampling volume location. This obstacle was exacerbated in acutely rejecting grafts, where the 2D image of the tissue myocardium gradually became blurry, making it difficult to discern consistent sampling locations.

3.8.3 Tricuspid Flow

Tricuspid Doppler inflow was easily acquired using UBM-Doppler, however, the small, variable, and inconsistent patterns seen in the biphasic waveforms of tricuspid flow did not yield any significant differences between the groups of grafts or any temporal changes within each experimental group (see **Appendix**).

3.9 Flow in the Ascending Aorta

A sample 2D image and a Doppler spectrum of mid-aortic blood flow velocity that shows derived flow parameters (i.e., peak velocities, time-velocity integrals, and heart rates) are shown in **Fig. 19A,** and **B**. Higher forward and backward aortic flow velocities are commonly seen in acutely rejecting allografts (**Fig. 20**). Periodic wave-like patterns were seen as a result of asynchronous native and graft heart rates. Maximum forward velocity (Fwd Vmax) was similar for both iso- and allografts from days 1-5 (**Fig. 21A**) but a significant diverging trend was seen from day 8 onwards (*$P=0.003$). No temporal changes in aortic maximum velocities were observed in isografts, which remained stable in the lower

~200-300 mm/s range, whereas time differences were seen in allografts between early (days 1-5) and late PTx figures (days 13 or 14, Δ, $P \leq 0.001$). Bwd Vmax (**Fig. 21B**) was identical for both groups of grafts on day 1. Generally higher and more variable Doppler velocities were seen in allografts, probably due to the falling out of rejected allografts from the data. Isografts were stable in the low range (~100-200 mm/s) and showed no significant instantaneous changes throughout the observed period. Statistical differences between isografts and allografts dominated the majority of the latter half of the experiment (days 6-13, *$P \leq 0.001$).

Time-velocity integrals (TVI) traces were performed for maximum peak velocities (Vmax) of forward and backward waveforms (**Fig. 19B**). Fwd TVI (**Fig. 22A**) for isografts and allografts was similar for both groups on days 1-5. Isograft Fwd TVI decreased stably afterwards with generally higher values seen in allografts (0.8-0.6 cm vs. 1.2-1.4 cm). Significant differences between iso- and allografts were detected on days 7, 8, 10 and 13 (*$P=0.004$). Similar to the pattern in both Fwd Vmax and Fwd TVI, Bwd TVI (**Fig. 22B**) was also low (<0.5 cm) and stable for isografts throughout the PTx period. Allograft Bwd Vmax diverged from isografts on day 7, achieving significantly higher values for the remainder of the observed period. Temporal changes within allografts were detected between early (1,3-5) vs. late-transplant values (days 13 or 14, Δ, $P \leq 0.001$). Plots for a modified Fwd RI and Fwd Ved are included in the **Appendix**.

3.10 *Coronary Perfusion*

3.10.1 *Left Coronary Artery Doppler Waveforms and Resolution Analysis*

Analysis of the complex coronary arterial Doppler waveform was carried out in a collaborative study (Zhou *et al.*, 2007). Similar to aortic flow, the periodic wave-like patterns were caused by asynchrony between native and graft HRs (**Fig. 23A, B**). Forward flow (**Fig. 23C**) was primarily responsible for perfusing the graft myocardium and was presumably due to native circulatory pressure. The waveform assumed to be produced primarily by the graft cardiac cycle is shown in **Fig. 23D**; positive (forward) flow occurs during ventricular relaxation and negative (backward) flow during isograft ventricular contraction. **Fig. 23E** shows the summation of the periodic waveforms of **Fig. 23C** and **D**, which closely approximates an actual Doppler spectrum for the LCA (**Fig. 23B**). The various vertical lines

65

in panels **C** to **E** refer to different stages of the cardiac cycles of the native and graft hearts that were correlated to the different components of the coronary waveform and are summarized in **Table VI**. A sample Doppler spectrum for acquiring Vmax, TVI, Ved, and native and graft HRs is shown (**Fig. 24A**). The serial images of **Fig. 24B** showed an appreciable finding - coronary waveforms of acutely rejecting allografts (**Fig. 24,** *ALLO:D10*) was exceptionally similar to the pulsatile and continuous flow seen in **Fig. 23C**, which shows flow due only to recipient circulatory pressure. This suggests that poor graft viability of a typical day 10 allograft yields a unique Doppler spectrum that lacks waveforms due to graft ventricular contraction, namely the absence of backward flow.

3.10.2 Left Coronary Arterial Flow Velocities and Waveform Integrals

Maximum forward velocity (Fwd Vmax, **Fig. 25A**) was similar for both iso- and allografts from days 1-4. Allograft Fwd Vmax peaked on day 5, which was significantly different from day 1 (Ψ, $P \leq 0.001$) and day 14 values (Δ, $P \leq 0.001$), and decreased gradually thereafter. Isografts values peaked on day 6, which also differed from day 1 PTx. Transient statistical differences between iso- and allografts were detected on days 5, 10 and 13 (*$P \leq 0.001$). Maximum backward velocity (Bwd Vmax, **Fig. 25B**) values showed a similar temporal pattern as Fwd Vmax for both groups of grafts from days 1-4. Allograft values peaked at day 6 and decreased thereafter, whereas isografts Bwd Vmax remained at high values and significantly different from allografts on days 11-13 (*$P \leq 0.001$). The peak on days 5-7 in allografts was significantly different from day 1 (Ψ, $P \leq 0.001$) and day 14 values (Δ, $P \leq 0.001$). Similar to AA measurements, TVI traces were performed for maximum peak velocities (Vmax) of forward and backward waveforms (**Fig. 26A, B**). Fwd TVI was similar for both iso- and allografts for the majority of the PTx period, with transient differences detected on days 5 and 9 only (*$P \leq 0.001$). Allograft TVI figures peaked early on day 5, which differed from day 1 values (Ψ, $P=0.016$). Bwd TVI (**Fig. 26B**) data shows a similar temporal pattern for both groups of grafts from days 1-4. Allografts peaked at day 6 and decreased thereafter. Isograft Bwd TVI diverged from allograft figures from day 9 onwards, though this trend was not statistically different. For allografts, days 4-7 differed from day 14 values (Δ, $P \leq 0.001$) and for isografts, day 5 figures were significantly higher than day 1 values (Ψ, $P=0.023$). Plots for a modified RI and Ved are included in the **Appendix**.

66

3.10.3 Right Coronary Arterial Flow

Similar analyses for LCA data were carried out for RCA data (**Fig. 27A, B**). As with the LCA, serial images of **Fig. 28B** show RCA waveforms in acutely rejecting grafts (**Fig. 28B**, *ALLO:D10*) to be exceptionally similar to the pulsatile and continuous flow seen in **Fig. 23C**, where flow is due only to native circulatory pressure. Fwd Vmax increased for both iso- and allografts from days 1-3 and allograft figures reached a maximum during days 4-6 and decreased gradually thereafter (**Fig. 29A**). Isografts figures remained at higher values close to ~500 mm/s after day 6. Significant differences were detected on days 8-10 and 12 (*$P \leq 0.001$*). Bwd Vmax increased for both groups of grafts early PTx but the pattern in allografts was variable, which peaked at day 2 and 4 and decreased to low values after day 7 (**Fig. 29B**). Isografts remained in the ~100-200 mm/s range after day 7. Significant differences were detected between isografts and allografts on days 9-12 and 14 (*$P \leq 0.001$*). Allograft Fwd TVI peaked on day 4, decreased to a minimum on day 9, and was variable afterwards (**Fig. 30A**). Fwd TVI for Isografts remained generally stable in the 2-3 cm range. Only transient significant differences were seen on days 8, 9 and 13 (*$P \leq 0.023$*). Bwd TVI (**Fig. 30AB**) increased for isografts from days 1-3 and generally remained stable at 0.4 cm. Allograft Bwd TVI values peaked early on day 2 and rapidly declined from day 7 onwards. Significant differences were seen between the two groups of grafts from day 9 onwards (*$P=0.001$*). Plots for a modified RI and Ved are included in the **Appendix**.

3.11 Main Pulmonary Arterial (MPA) Flow

MPA Doppler flow was acquired using UBM-Doppler from a SAV view of the cardiac base, showing the right ventricular outflow tract. However, measurements of MPA Vmax and TVI showed inconsistent patterns and did not yield any significant differences between the groups of grafts or any temporal changes within each experimental group (see **Appendix**).

3.12 Pericardial Effusion, Cardiac Dilation, and Aortic Thrombi

Fluid surrounding the graft was present to some degree in all test subjects. Though the actual volume of pericardial (or perigraft) effusion was not quantified because of limitations of our approach, qualitative observations indicate that effusion occurred to a much greater degree in allografts when compared to isografts and was more likely to occur late after transplant (**Fig. 31A**). Dilation of the LA occurred in 7 of the 8 isografts and in none of the allografts used for the study and was therefore, a phenomenon attributed exclusively to syngeneic transplants (**Fig. 31B**). This phenomenon was not quantified due to the difficulty of determining exact atrial chamber wall dimensions, but was generally found to occur in association with high MR values. In allografts, LV remodelling was observed (**Fig. 31 C, D**) to induce increases in chamber dimensions late after transplant (**Fig. 14A, B**). It is important to note that the observation of increased LV dimensions was not only ubiquitous in all acutely rejecting grafts, but seemed independent of the day PTx but dependent on the day of rejection for that particular graft. For example, one allograft rejected on day 7 and showed significantly increased LV dimensions, whereas another allograft that rejected on day 17 did not develop LV dilation until after day 10. The evidence from B-mode imaging of LV dilation corroborates with the quantified increased end-systolic and end-diastolic dimensions seen in allografts. Moreover, though the formation of LV thrombi was common in all grafts, in 2 isografts and 3 allografts, ventricular thrombus dislodged from the LV chamber and traversed the aortic orifice and into the AA, alternating locations between the LV and AA during diastole and systole, respectively (**Fig. 31 E, F**). This phenomenon produced atypical aortic and coronary Doppler waveforms as well as an observed increase in graft HR and higher palpation scores (data not shown). The presence of aortic thrombi may be a common occurrence not previously documented in this model of heterotopic transplantation and may be a factor in artificially increasing the force of palpable cardiac contractility.

3.13 Histology and TUNEL Analyses

Rejected allograft specimens showed extensive and diffuse mononuclear cell infiltration associated with patchy regions of myocyte destruction and interstitial edema (**Fig. 32D, E**); histopathology was absent in isografts (**Fig. 32A, B**). Areas of considerable destruction in allografts were visible next to mildly affected areas within a single graft, consistent with the patchy rejection process. In contrast to isografts, TUNEL-positive cells were easily identifiable in rejected allografts and randomly distributed within the myocardium (**Fig. 32F**). In isografts, however, TUNEL-positive cells were absent (**Fig. 32C**).

3.14 Data Summary and Interactions

The data for representative days post-transplant (1, 3, 6 and 10) are summarized in **Table VII.** Interactions terms and p-values are tabulated in **Table VIII.**

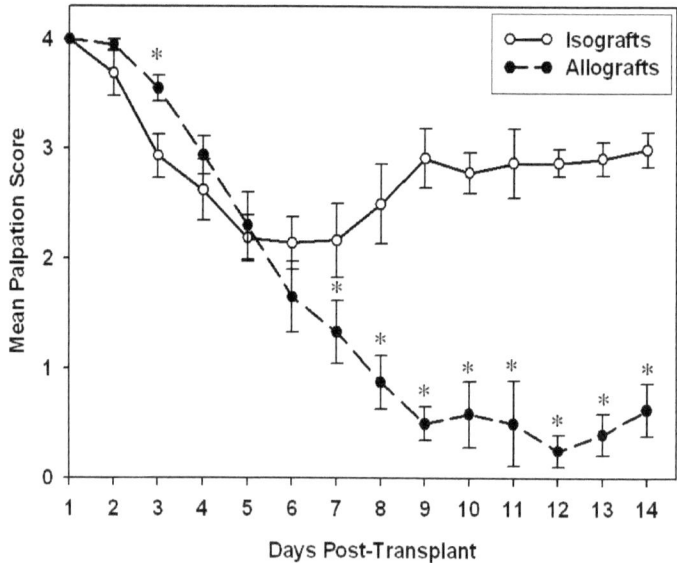

Allografts (n) = (10) (10) (10) (10) (10) (9) (8) (8) (6) (6) (5) (5) (4) (4)

Figure 3 – Mean Palpation Scores in Isografts and Allografts. Palpation scores were assigned for intensity of graft pulsation. Serial assessments for allografts were taken until the day of rejection as assessed by echocardiographic determination (see methods); isografts were scored daily for 14 days. A scale from 0 to 4 was employed; 4 indicates rigorous cardiac activity whereas 0 indicates cessation of activity. The figures in brackets are identical for all plots and indicate the number of surviving allografts for that day. Seven (7) isografts survived for the full period of observation. Data are reported as mean ± SEM. *P = <0.001 between isografts and allografts.

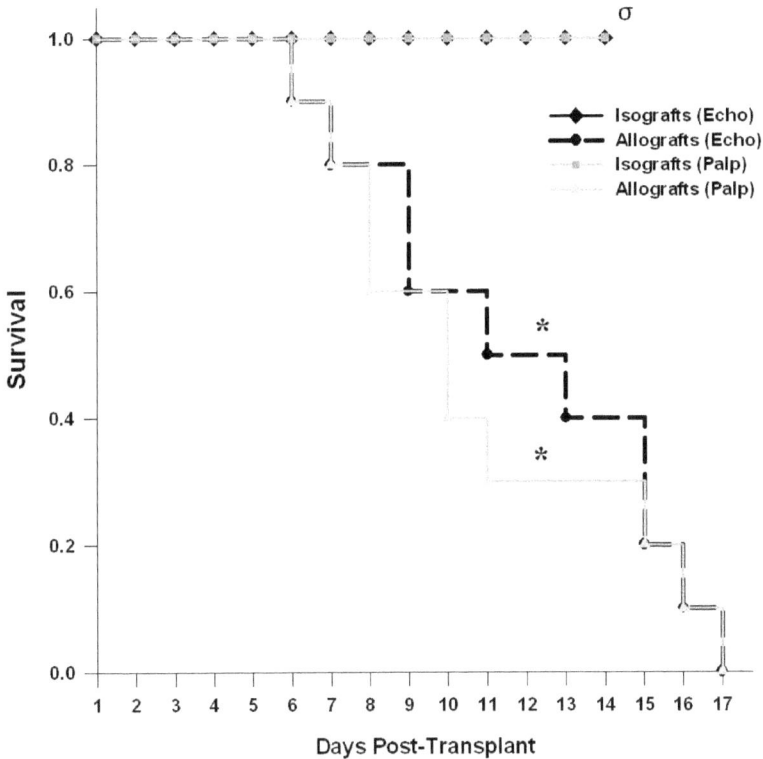

Figure 4 – Graft Survival by Palpation and Echocardiography. Differences in graft survival were detected between isografts and allografts by palpation (*P = 0.002) and echocardiography (*P = 0.003) but graft survival did not significantly differ between methods. Median survival time (MST) by palpation (10 days) and echocardiography (12 days) was not significantly different. Isografts were sacrificed at 14 days (σ) but were presumed to survive past 17 days in this study.

Table V. Allograft Survival and Median Survival Time (MST) by Palpation and Echocardiography

Donor	Recipient	N	Survival by palpation (days)	MST (days)	*P	**Survival by echocardiographic determination (days)	MST (days)	*P
C3H/He	C3H/He	7	*Sacrificed at 14 days post-transplant*	–		*Sacrificed at 14 days post-transplant*	–	
BALB/c	C3H/He	10	6, 7, 8, 8, 10, 11, 15, 16, 17	10	0.002	6, 7, 9, 9, 11, 13, 15, 15, 16, 17	12	0.003

*MST = Median survival time; *P value of significance between isografts and allografts; **See methods for criteria*

71

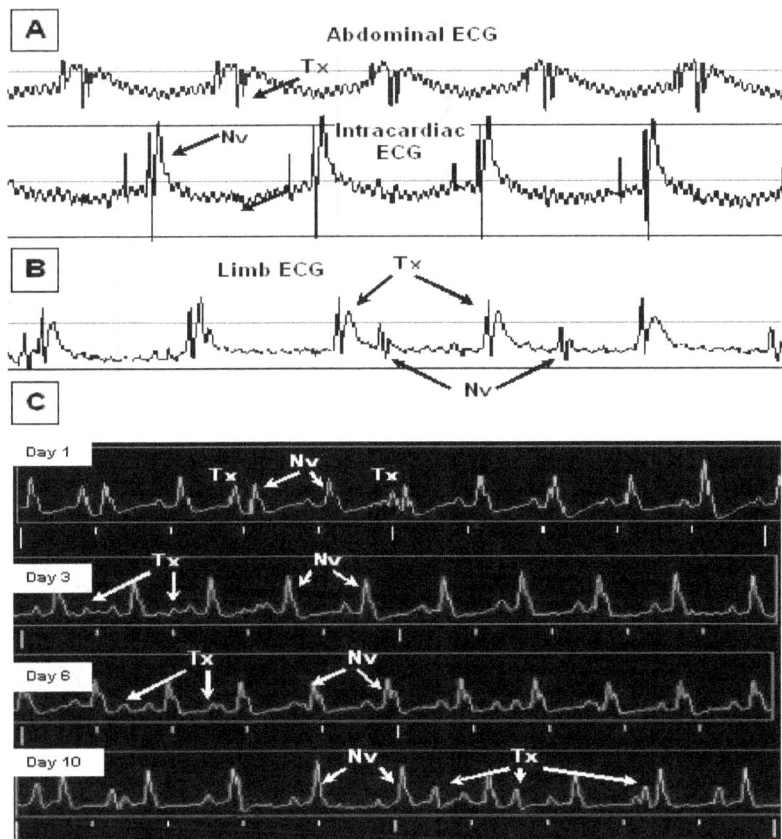

Figure 5 – Cardiac Graft ECG. A. Abdominal ECG and simultaneous intracardiac (native heart) ECG trace in an anaesthetized isografted mouse. B. Limb ECG obtained from the same isografted mouse as in A at a different time (i.e., non-simultaneous). C. UBM limb lead ECG traces for an anaesthetized allografted mouse on days 1, 3, 6, and 10 post-transplant. Nv, ECG waveforms caused by the native heart; Tx, ECG waveforms caused by the cardiac graft.

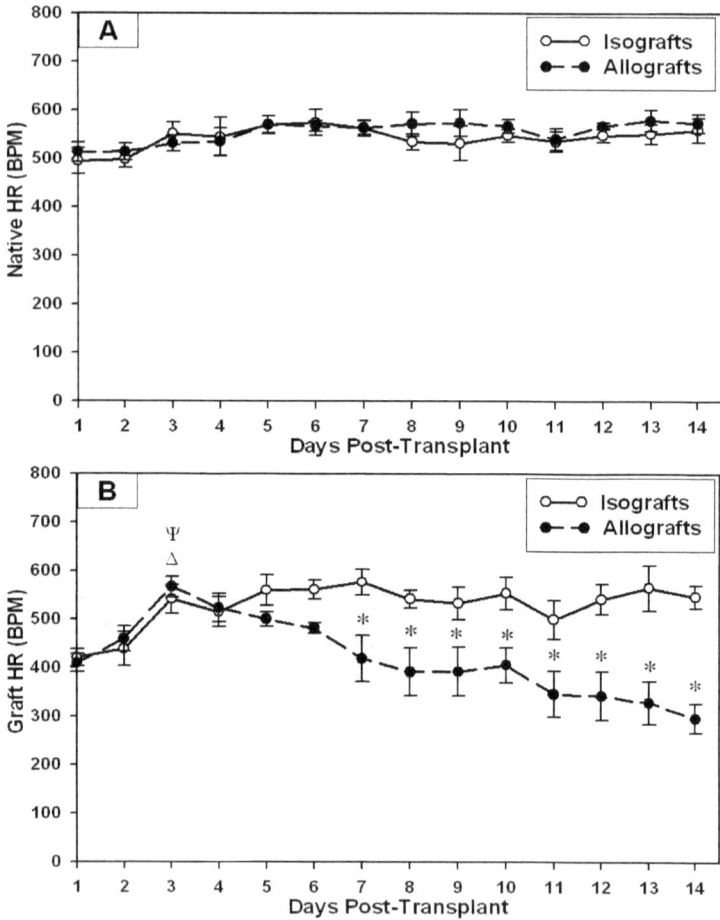

Figure 6 – Native and Graft Heart Rates. Heart rates were obtained from aortic Doppler spectra and supplemented with LCA-Doppler data (<2% of total data) when necessary. **A.** No differences in native HR were detected between groups of grafts or over time. **B.** Differences in graft HR were detected between isografts and allografts beginning on day 7 onwards (*$P \leq 0.001$). (Δ) indicates significance vs. day 14 for allografts only ($P \leq 0.001$) and (Ψ) for differences vs. day 1 for both isografts and allografts ($P \leq 0.001$). Data are reported as mean ± SEM.

73

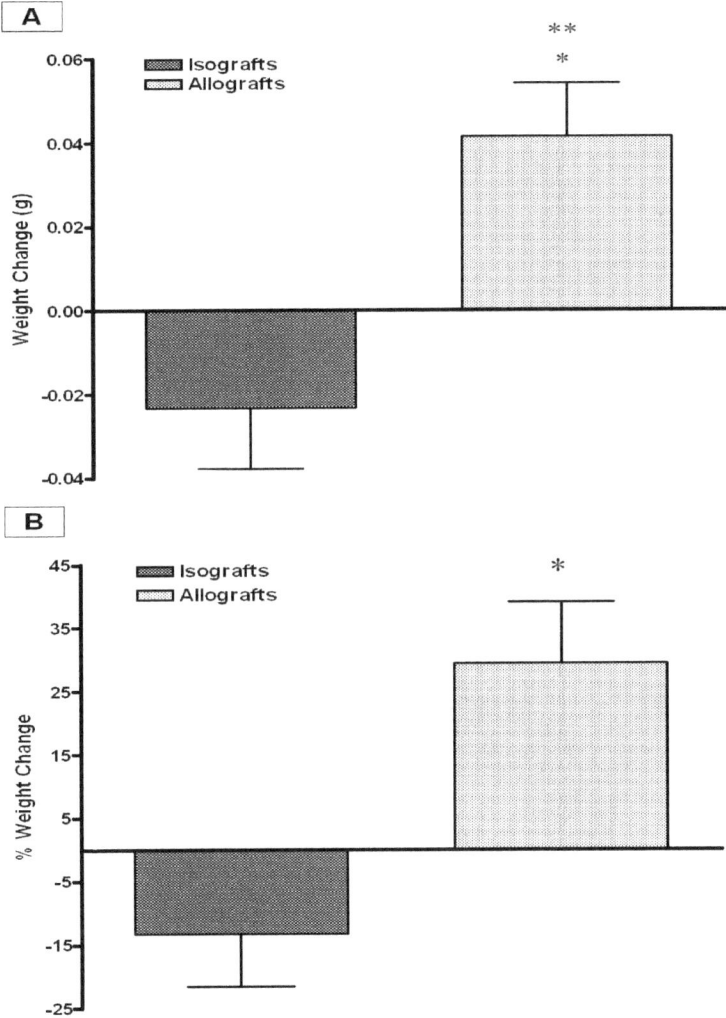

Figure 7 – Absolute and % Graft Weight Changes before Implantation and after Explantation. The change in myocardial mass of rejected allografts (n=5) was significantly greater than in isografts (n=5). Allografts were explanted at rejection (MST=12 days) and isografts at 14 days post-transplant (raw weight change, *P = 0.010; % weight change, *P = 0.009). Allograft post-explantation weight was significantly higher than pre-implantation weight (**P = 0.031). Data are reported as mean ± SEM.

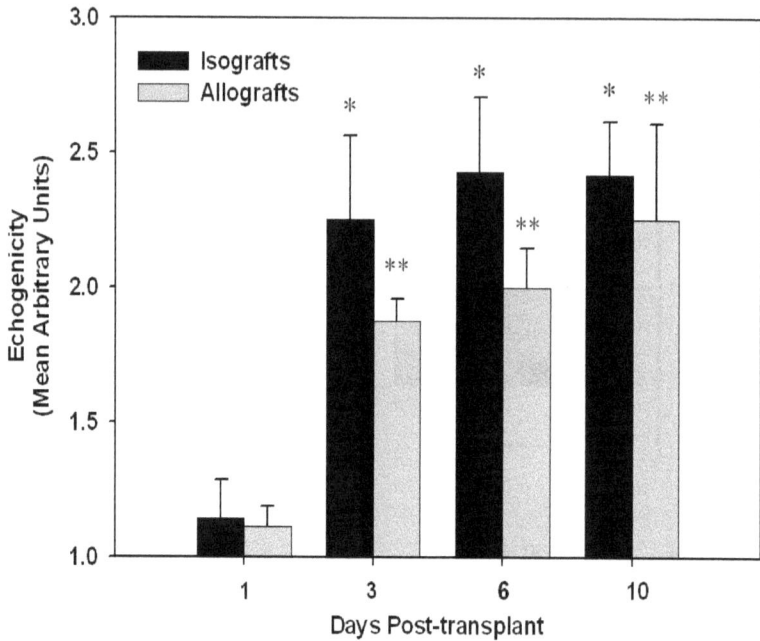

Figure 8 – Semi-Quantitative Assessment of Echogenicity. Backscatter intensity was scored semi-qualitatively in a blinded fashion on a scale from 1 (low intensity) to 3 (high intensity). Differences in myocardial backscatter intensity were detected in isografts (*$P \leq 0.001$) and allografts (**$P \leq 0.001$); both show significant intensity increases after day 1 post-transplant. No differences were detected between isografts vs. allografts in each respective day. Data are reported as mean ± SEM.

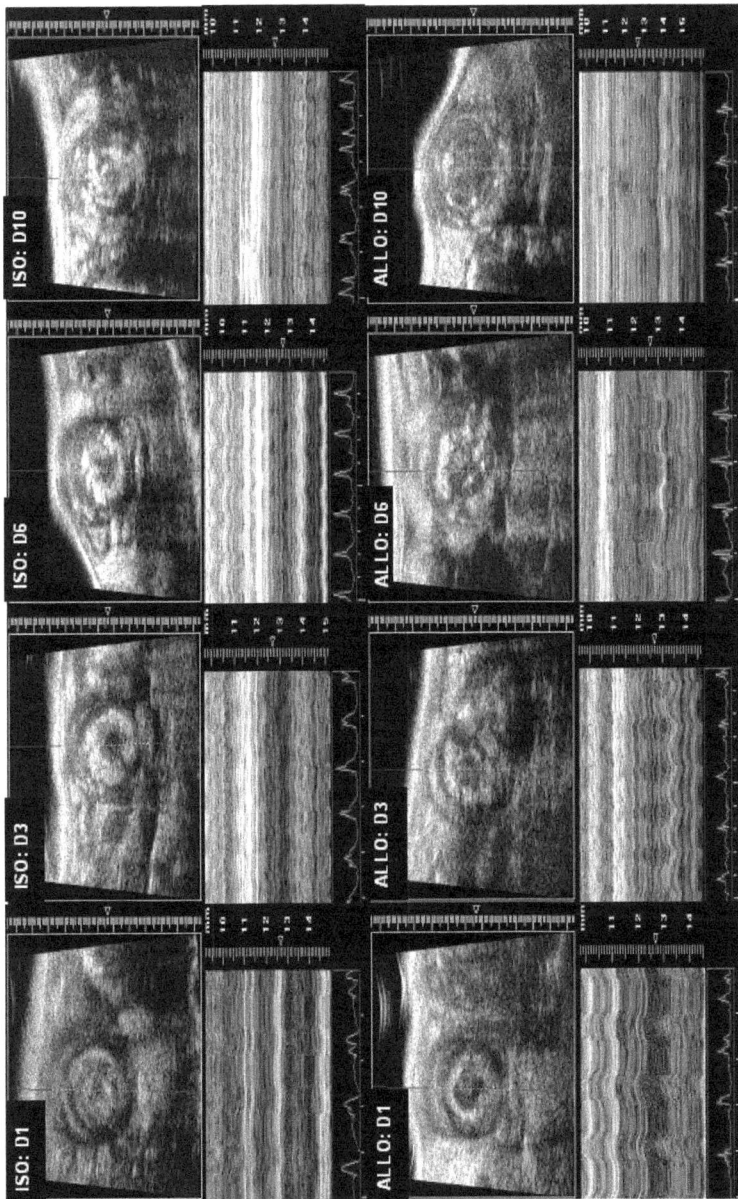

Figure 9 – Serial B-mode and Corresponding M-mode Short-Axis View (SAV) Images. Two-dimensional B-mode images of the LV in the short-axis view (SAV) on days 1, 3, 6, and 10 post-transplant in isografts (ISO) and allografts (ALLO) with corresponding M-mode spectra and ECG. Placement of the M-mode cursor line was placed at the largest LV chamber dimension.

76

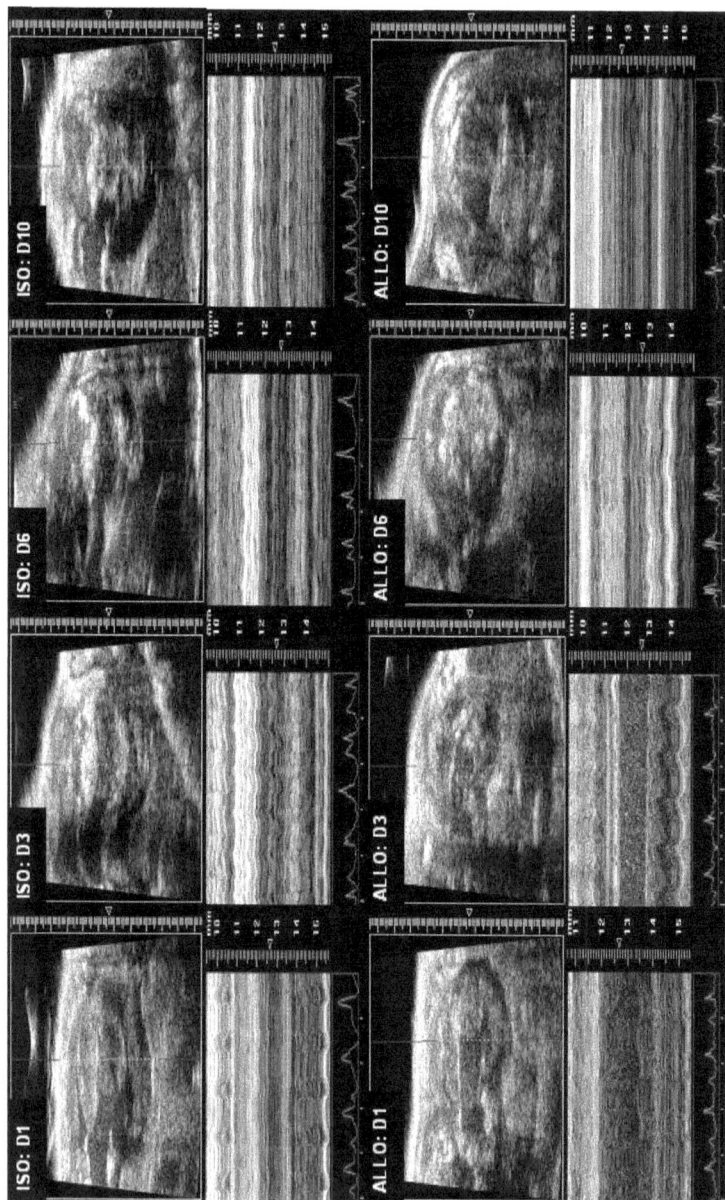

Figure 10 — Serial 2-Dimensional Long-Axis View (LAV) images and Corresponding M-mode Spectra. Two-dimensional B-mode images of LV in the long-axis view (LAV) on days 1, 3, 6, and 10 post-transplant in isografts (ISO) and allografts (ALLO) with corresponding M-mode spectra and ECG. The M-mode cursor line was placed on the section of the LV that yielded that largest chamber dimension.

77

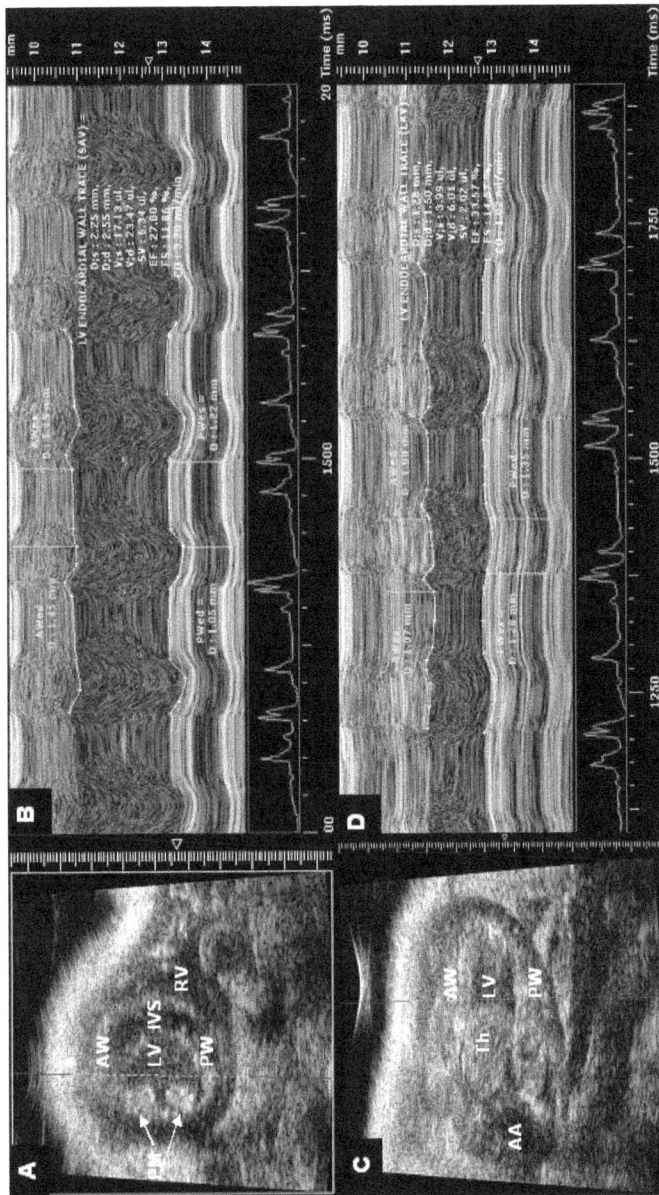

Figure 11: Acquisition of Physiologic Parameters of Graft Function from Left Ventricle M-mode Spectra. Representative 2-D B-mode images of the left ventricle are shown on the left panels in the short-axis (SAV, **A**) and long-axis (LAV, **C**) views; representative M-mode spectra and ECG are in the right panels (**B**, **D**). M-mode cursor lines are positioned to measure largest LV dimensions. Automated analysis of chamber dimensions and physiologic parameters (%EF, %FS, CO, SV) were obtained from semi-automated traces of the LV endocardial wall. *LV*, left ventricle; *RV*, right ventricle; *AW*, anterior wall; *PW*, posterior wall; *PM*, papillary muscles; *IVS*, interventricular septum; *Th*, ventricular thrombus; *AA*, ascending aorta; *Awes*, anterior wall thickness at end-systole; *Awed*, anterior wall thickness at end-diastole; *D.s*, end-systolic diameter; *D.d*, end-diastolic diameter; *PWed*, posterior wall thickness at end-systole; *PWes*, posterior wall thickness at end-diastole

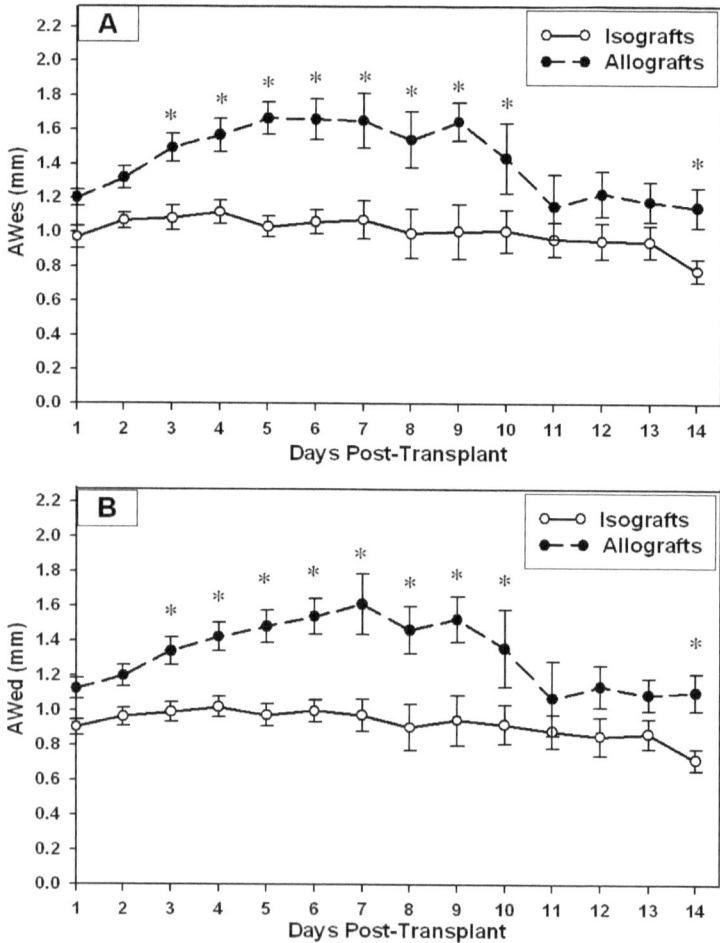

Figure 12 – End-Systolic (AWes) and End-Diastolic (AWed) Left Ventricular Anterior Wall Thicknesses. M-mode images were averaged from both the SAV and LAV in isografts and allografts. Differences in anterior wall thicknesses in allografts vs. isografts were seen from days 3 to 10 and 14 in AWes (**A**) and AWed (**B**) (*$P \leq 0.001$). Data are reported as mean ± SEM.

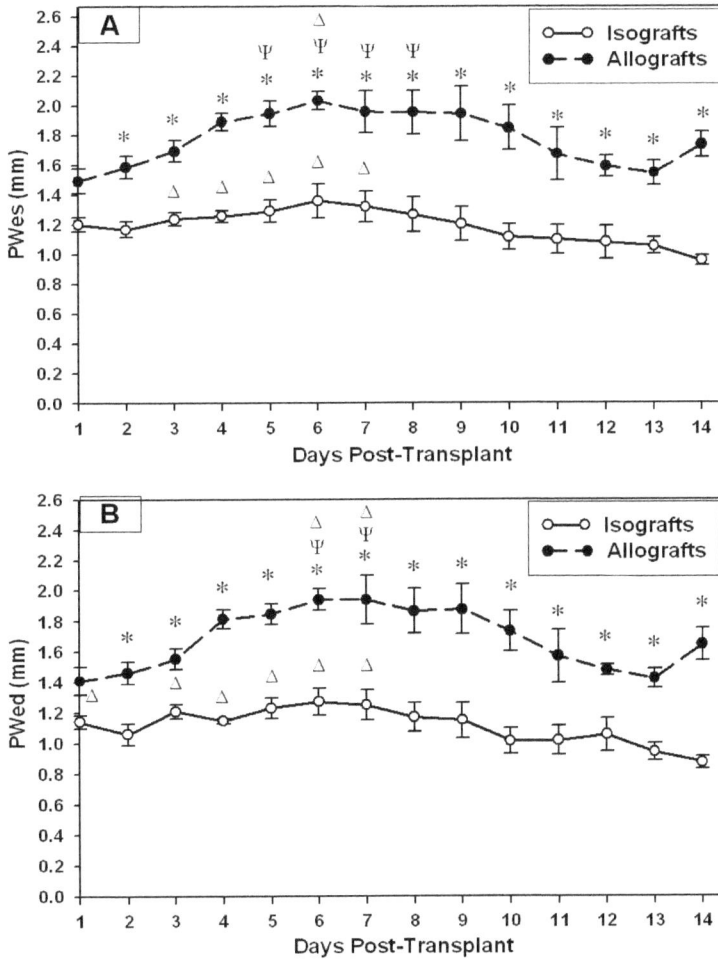

Figure 13 – End-Systolic (PWes) and End-Diastolic (PWed) Left Ventricular Posterior Wall Thicknesses. Differences in posterior wall thicknesses in allografts vs. isografts were seen from day 2 onwards for PWes (**A**) and PWed (**B**) (*$P \leq 0.001$). (Ψ, $P \leq 0.001$) indicates significant differences vs. day 1 and (Δ, $P \leq 0.001$) indicates significance vs. day 13 or 14. Data are reported as mean ± SEM.

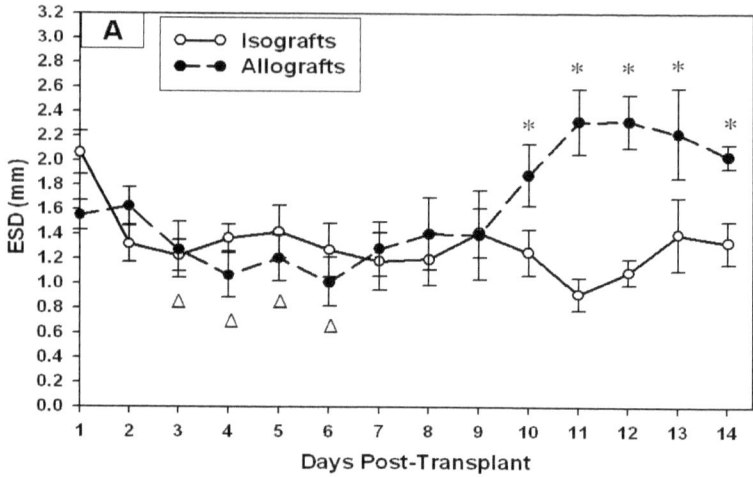

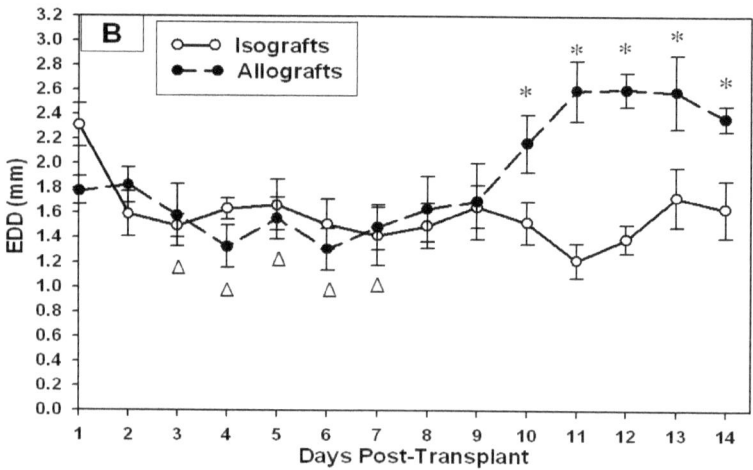

Figure 14 – End-Systolic (ESD) and End-Diastolic (EDD) Left Ventricular Diameters. Allograft left ventricular ESD (**A**) and EDD (**B**) increased significantly in comparison to isografts beginning on day 10 and onwards. (*$P \leq 0.001$). Allograft ESD and EDD reached peak values late after transplant. ($\triangle$, $P \leq 0.001$) reflects significance vs. day 13 or 14 for allografts only. Data are reported as mean ± SEM.

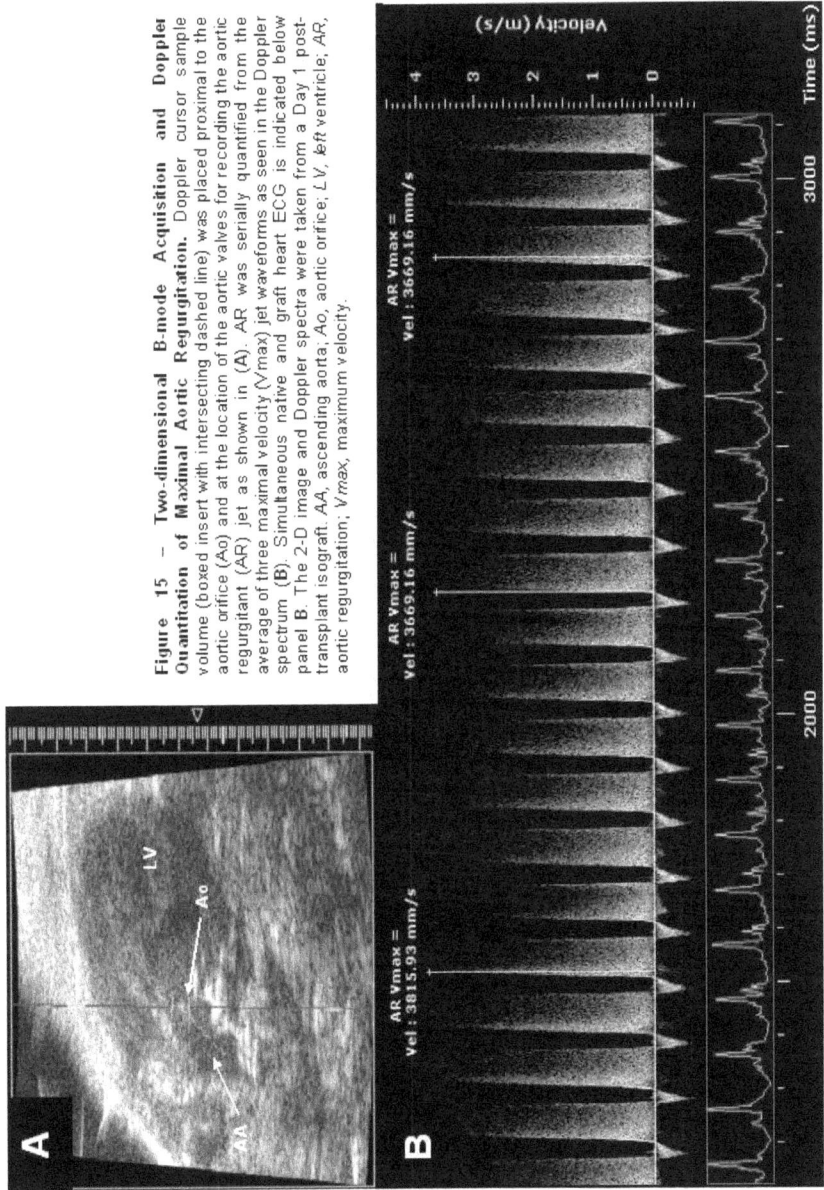

Figure 15 – Two-dimensional B-mode Acquisition and Doppler Quantitation of Maximal Aortic Regurgitation. Doppler cursor sample volume (boxed insert with intersecting dashed line) was placed proximal to the aortic orifice (Ao) and at the location of the aortic valves for recording the aortic regurgitant (AR) jet as shown in (A). AR was serially quantified from the average of three maximal velocity (Vmax) jet waveforms as seen in the Doppler spectrum (B). Simultaneous native and graft heart ECG is indicated below panel B. The 2-D image and Doppler spectra were taken from a Day 1 post-transplant isograft. AA, ascending aorta; Ao, aortic orifice; LV, left ventricle; AR, aortic regurgitation; Vmax, maximum velocity.

82

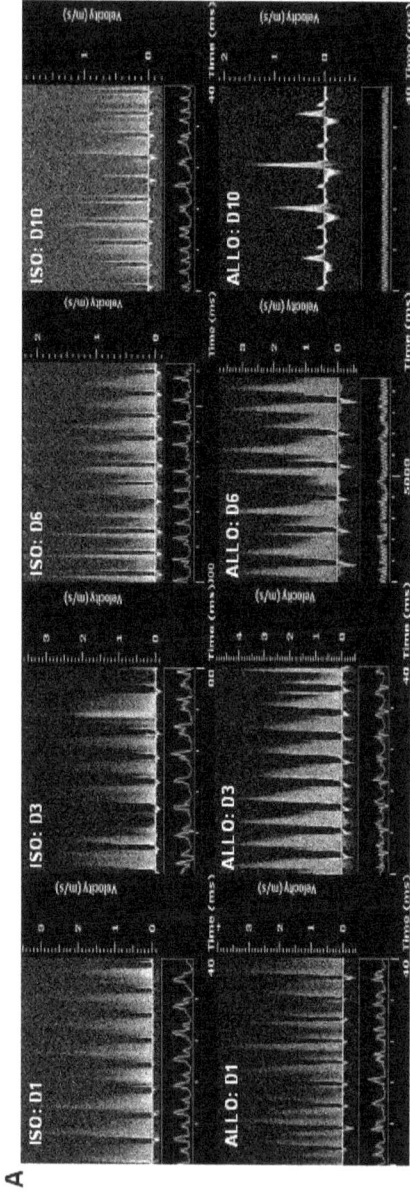

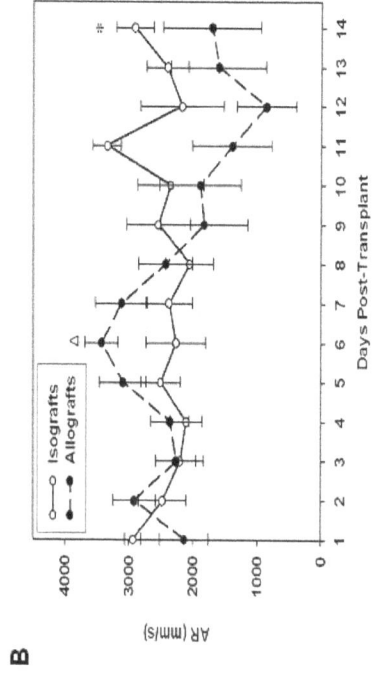

Figure 16 - Serial Doppler Characterization of Aortic Regurgitation (AR) in Isografts and Allografts. A. Doppler spectra produced by the high velocity regurgitant jet at the aortic orifice on days 1, 3, 6, and 10 post-transplant in isografts (ISO) and allografts (ALLO) with ECG. **B.** Doppler waveforms due to the regurgitant jet was averaged from three cardiac cycles to obtain mean AR. Differences between isografts and allografts were detected on day 14 only (*P=0.023). (Δ) indicates significant differences vs. late-transplant values in allografts (days 12-14, P=0.001). Data are reported as mean ± SEM.

83

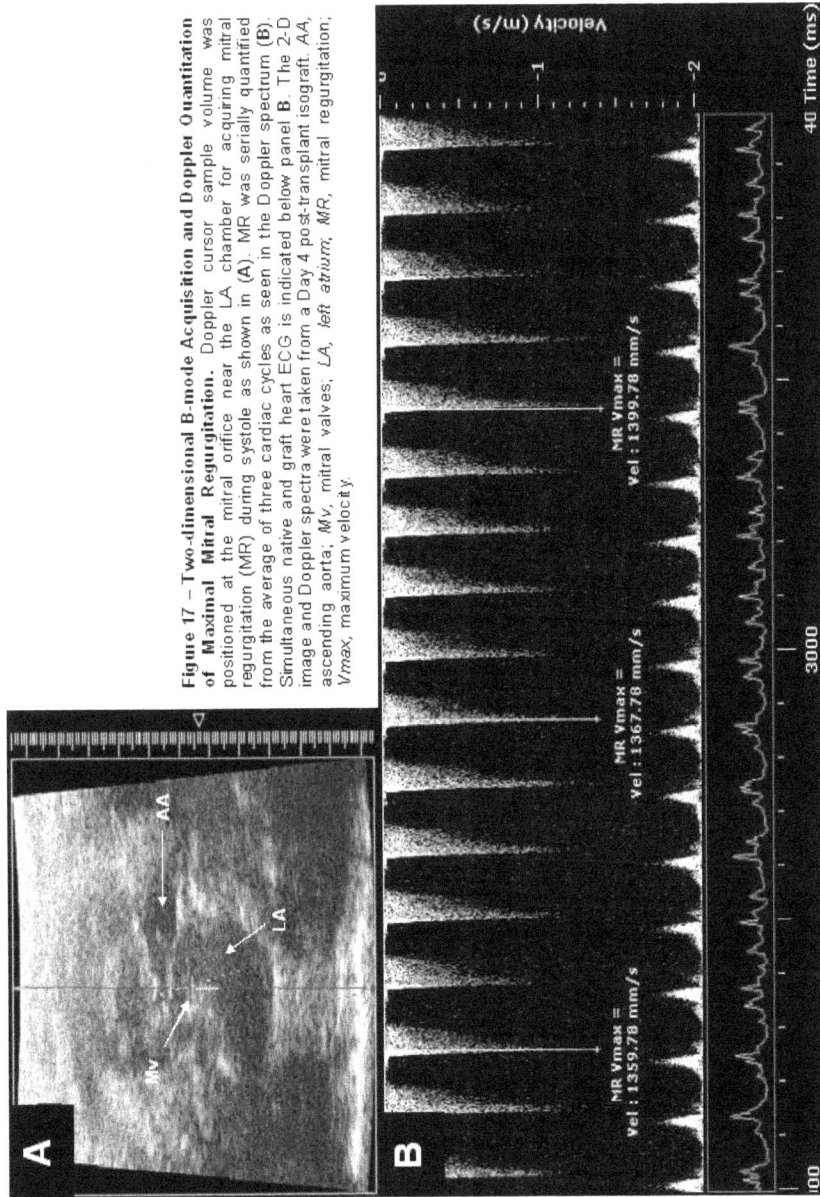

Figure 17 – Two-dimensional B-mode Acquisition and Doppler Quantitation of Maximal Mitral Regurgitation. Doppler cursor sample volume was positioned at the mitral orifice near the LA chamber for acquiring mitral regurgitation (MR) during systole as shown in (**A**). MR was serially quantified from the average of three cardiac cycles as seen in the Doppler spectrum (**B**). Simultaneous native and graft heart ECG is indicated below panel **B**. The 2-D image and Doppler spectra were taken from a Day 4 post-transplant isograft. *AA*, ascending aorta; *Mv*, mitral valves; *LA*, left atrium; *MR*, mitral regurgitation; *Vmax*, maximum velocity.

84

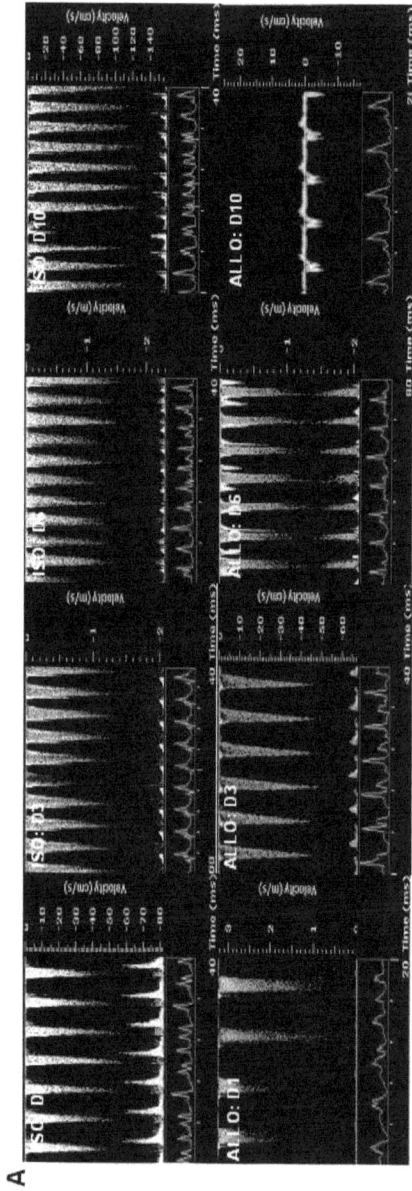

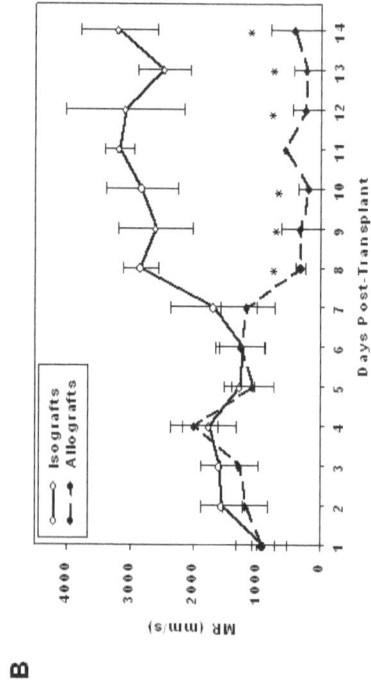

Figure 18 - Serial Doppler Characterization of Mitral Regurgitation (MR) in Isografts and Allografts. A. Doppler spectra produced by the high velocity regurgitant jet at the mitral valve on days 1, 3, 6, and 10 post-transplant in isografts (ISO) and allografts (ALLO) with ECG. **B.** Doppler waveforms due to the regurgitant mitral jet were averaged from three cardiac cycles to obtain mean MR. *P≤0.001 between iso- and allografts. Data are reported as mean + SEM.

85

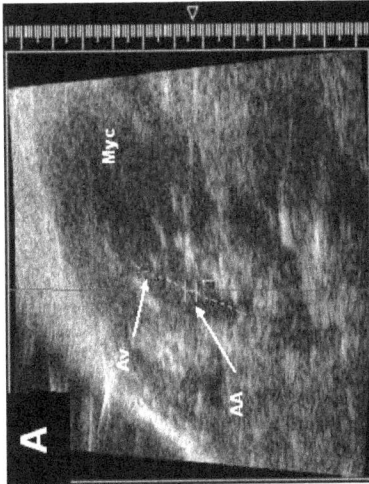

Figure 19 – Two-dimensional B-mode Acquisition and Doppler Quantification of Aortic Doppler Flow. Doppler cursor sample volume was positioned at the mid region of the ascending aorta (AA) to obtain a Doppler flow spectrum (A). Native HR was measured from the peaks of four consecutive positive waveforms (due to native pressure), whereas graft HR was determined from the peaks of four consecutive negative waveforms (due to graft systole) (B). The highest positive waveform constituted the maximum forward velocity (*Forward Vmax*) and the highest negative waveform yielded the maximal backward velocity (*Backward Vmax*). Time-velocity integrals (*TVI*) were measured for the highest forward and backward waveforms. *Ved* was also determined and *RI* calculated (see Methods). Simultaneous native and graft heart ECG is indicated below panel B. The 2-D image and Doppler spectra were taken from a day 2 isograft. *AA*, ascending aorta; *Av*, aortic valves; *Myc*, ventricular myocardium; *Vmax*, maximum velocity; *TVI*, time-velocity integral; *Ved*, forward end-diastolic velocity.

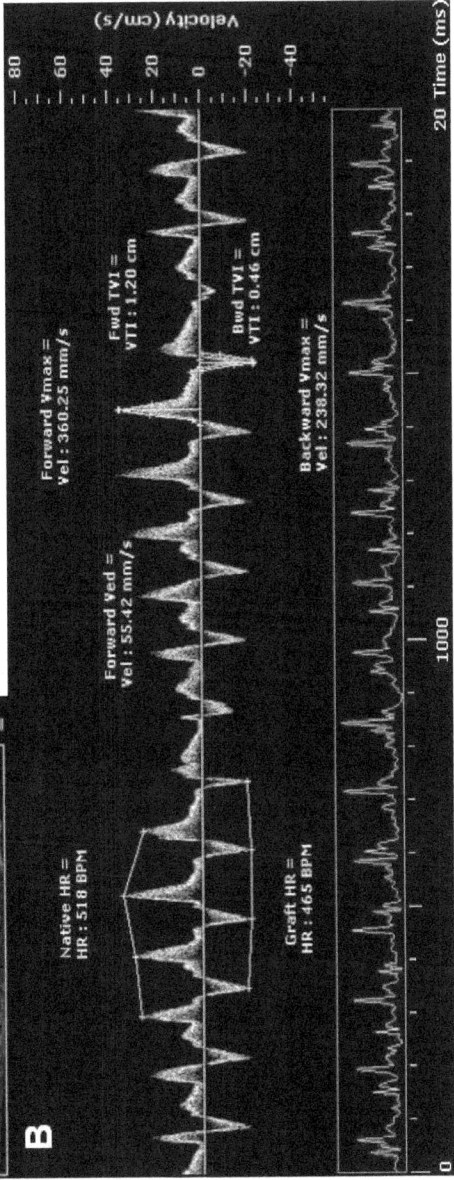

86

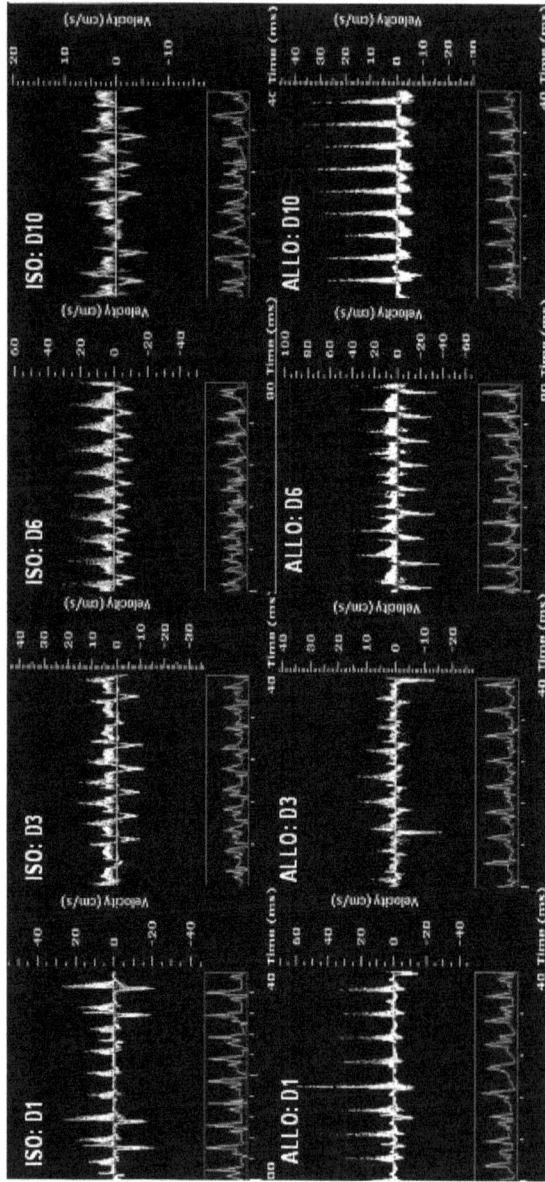

Figure 20 - Serial Doppler Characterization of Aortic flow in Isografts and Allografts. A. Doppler spectra produced by the asynchronous combination of native and graft heart rates, producing periodic fluctuations. These wave-like patterns were corroborated by the respective ECG traces in the lower panels, which show a similar pattern. Forward waveforms act to perfuse the myocardium whereas graft systole produces backward waveforms. Aortic flow is represented as a typical time-course: the immediate post-operative (day 1), recovery from ischaemic injury (day 3), mid (day 6), and late (day 10) post-transplant period. ISO, isografts; ALLO, allografts.

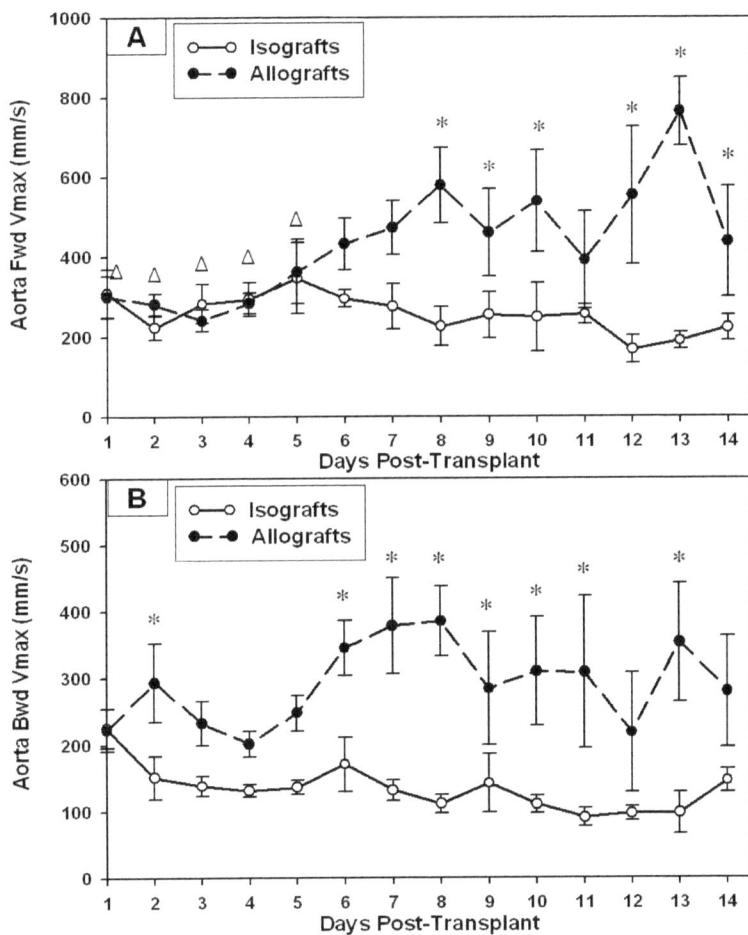

Figure 21 – Peak Aortic Forward and Backward Doppler Velocities.
A. Maximum forward velocity (Fwd Vmax, *P=0.003) and **B.** maximum backward velocity (Bwd Vmax, *P≤0.001) in the mid-aortic region. (*) indicates significance between iso- and allografts. (Δ) indicates significance vs. late transplant values for allografts only (day 13 or 14, P≤0.001). Data are reported as mean ± SEM.

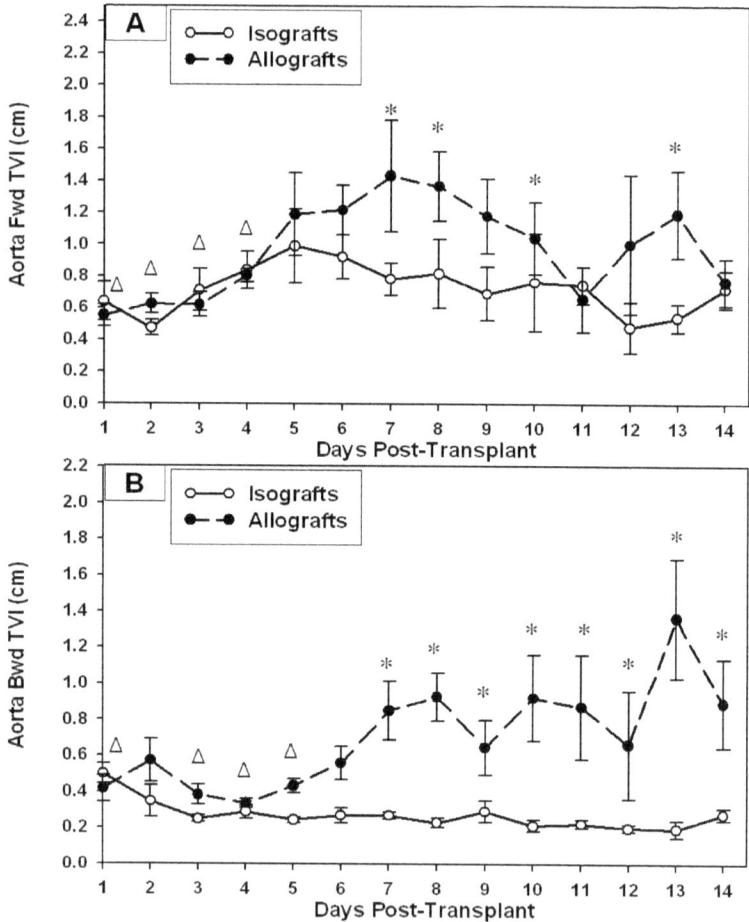

Figure 22 – Aortic Forward and Backward Time-Velocity Integrals (TVI). TVI traces were performed for maximum peak velocities (Vmax) of forward and backward waveforms. **A.** Time-velocity integrals for the maximum forward velocity (Fwd TVI, *P=0.004) and **B.** maximum backward velocity (Bwd TVI, *$P \leq 0.001$) in the mid-aortic region. (*) indicates significance between iso- and allografts. (Δ) indicates significance vs. late transplant values (day 13 or 14, $P \leq 0.001$) in allografts only. Data are reported as mean ± SEM.

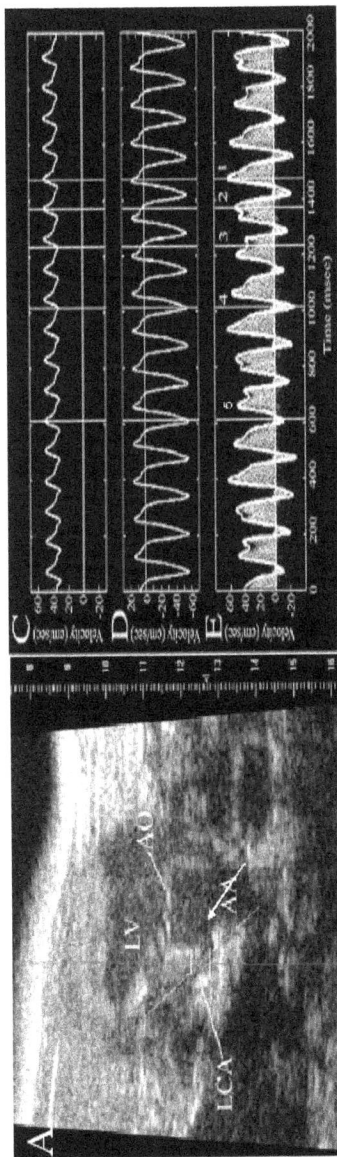

Table VI. Summary Table of Waveform Components

Parameter	Cardiac Cycle		Graft Flow Velocity	Vertical Line	Legend
	Native	Graft			
Fwl Vmax	ES	ED	-ve	1	a
*Native Vmax	ES	eS	~0	2	b
Fwl Ved	ED	eS	~0	3	c
Bwl Vmax	ED	ES	+ve	4	d
N/A	ES	ES	+ve	5	e

*Difficult to identify and exempt from further analysis

ES, end-systole; ED, end-diastole; eS, early systole. -ve, negative; +ve, positive

Figure 23 – UBM and Waveform Resolution of Coronary Doppler Spectra. Doppler cursor sample volume was positioned <1 mm distal to the origin of the left coronary artery (LCA) from the aortic sinus to obtain a Doppler spectrum (**A, B**). The Doppler spectrum shows the 'up direction' as forward (native arterial) flow and the 'down direction' as backward (graft systolic) flow. Forward flow (**C**) perfusing the graft myocardium approximated peripheral arterial flow with low resistance. (**D**) Presumed to be due to the graft cardiac cycle: positive (forward) flow corresponded to graft diastole and negative (backward) to graft systole. (**E**) Summing (**C**) and (**D**) closely approximates the actual Doppler spectrum shown in (**B**). In panels **C** to **E**, *line 1* corresponds to peak native heart systole and maximal graft diastole (Fwd Vmax) as indicated by *a* in panel (**B**). *Line 2* corresponds to peak native heart systole and simultaneous zero graft flow velocity (native Vmax, *b*). *Line 3* indicates end-diastole of the native heart with zero graft flow velocity (Fwd Ved, *c*). *Line 4* indicates graft systole at native heart end-diastole (backward Vmax, *d*). *Line 5* corresponds to both native and graft end-systole (*e*). **Table VI** summarizes the waveform components. The 2-D image and Doppler spectrum was taken from a day 4 and day 50 isograft, respectively. *AA*, ascending aorta; *LCA*, left coronary artery; *LV*, left ventricle; *Fwd*, forward; *Bwd*, backward; *Vmax*, maximum velocity; *Ved*, forward end-diastolic velocity. Panels A-E adapted from *Ultrasound Med Biol*, 33(**6**), Zhou *et al.*, *Morphological and functional evaluation of murine heterotopic cardiac grafts using ultrasound biomicroscopy*, pp. 870-9 (2007) with permission from Elsevier.

90

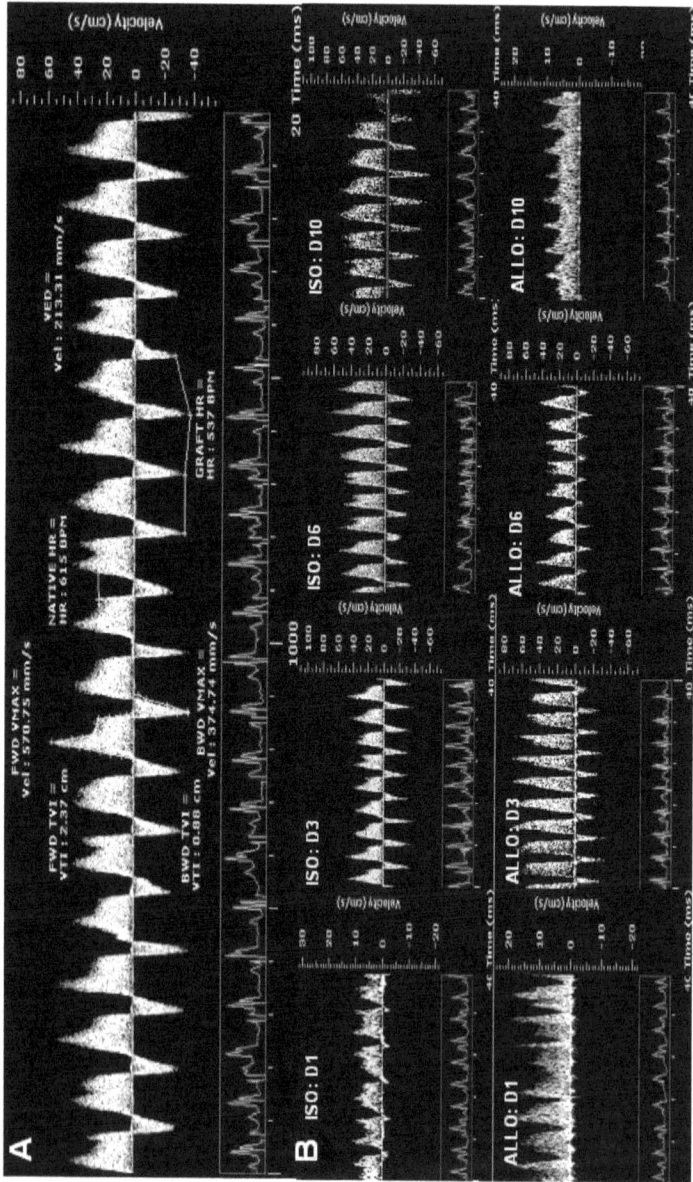

Figure 24 – Sample and Serial Left Coronary Artery (LCA) Doppler Spectra. A. Maximum velocities, end-diastolic velocity, time-velocity integrals, and heart rates were obtained from LCA Doppler spectra. The sample Doppler spectrum was taken from a day 4 isograft. **B.** Serial comparison of changes in LCA waveforms between isografts and allografts. Simultaneous native and graft heart ECG is indicated below panel B. *Fwd*, forward; *Bwd*, backward; *Vmax*, maximum velocity; *TVI*, time-velocity integral; *Ved*, forward end-diastolic velocity; *HR*, heart rate; *ISO*, isografts; *ALLO*, allografts.

91

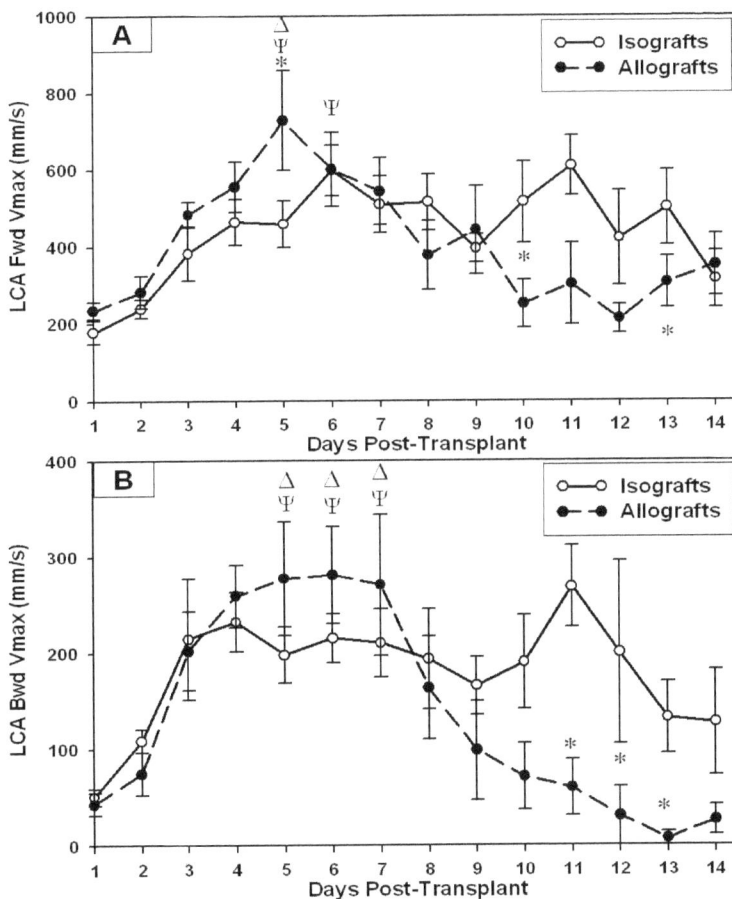

Figure 25 – Peak Left Coronary Artery Forward and Backward Doppler Velocities in Isografts and Allografts. A. Maximum forward velocity (Fwd Vmax). Differences between isografts and allografts are indicated as (*), vs. day 1 as (Ψ, allografts day 5, isografts day 6) and vs. day 13 or 14 values by (Δ). **B.** Maximum backward velocity (Bwd Vmax). Differences between isografts and allografts are shown as (*); for allografts, differences vs. day 1 (Ψ) and day 13 or 14 values (Δ) were detected. Significant differences are indicated at $P \leq 0.001$. Data are reported as mean ± SEM.

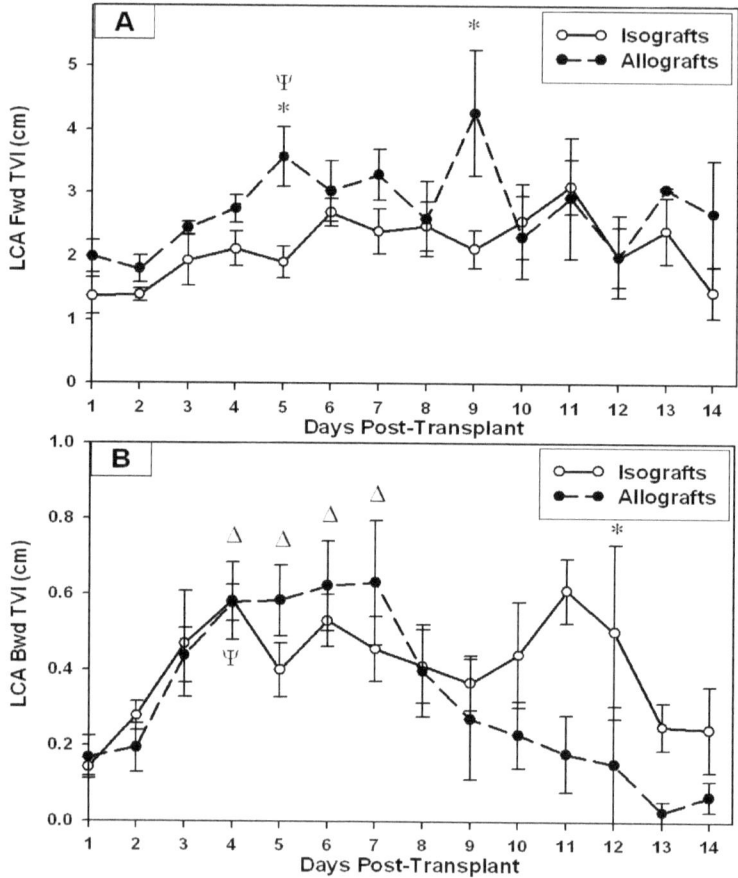

Figure 26 – Forward and Backward Time-Velocity Integrals (TVI) in the Left Coronary Artery. A. Maximum forward time-velocity integral (Fwd TVI). Differences between isografts and allografts are indicated as (*, $P \leq 0.001$); for allografts, differences vs. day 1 as (Ψ, $P=0.016$). B. Maximum backward TVI (Bwd TVI). Differences between isografts and allografts are indicated as (*, $P \leq 0.001$); for allografts, differences vs. day 13 or 14 values (Δ, $P \leq 0.001$) were detected and for isografts, differences vs. day 1 values (Ψ, $P=0.023$). Data are reported as mean ± SEM.

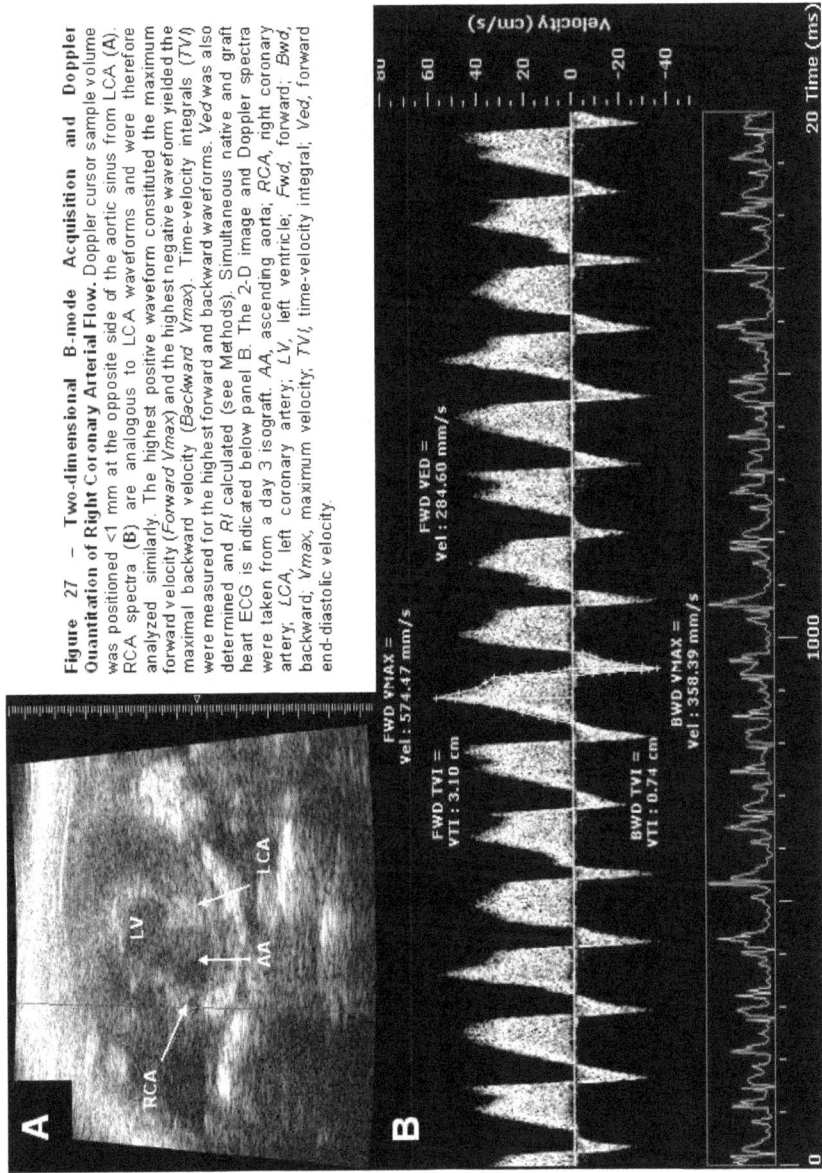

Figure 27 – Two-dimensional B-mode Acquisition and Doppler Quantitation of Right Coronary Arterial Flow. Doppler cursor sample volume was positioned <1 mm at the opposite side of the aortic sinus from LCA (**A**). RCA spectra (**B**) are analogous to LCA waveforms and were therefore analyzed similarly. The highest positive waveform constituted the maximum forward velocity (*Forward Vmax*) and the highest negative waveform yielded the maximal backward velocity (*Backward Vmax*). Time-velocity integrals (*TVI*) were measured for the highest forward and backward waveforms. *Ved* was also determined and *RI* calculated (see Methods). Simultaneous native and graft heart ECG is indicated below panel B. The 2-D image and Doppler spectra were taken from a day 3 isograft. *AA*, ascending aorta; *RCA*, right coronary artery; *LCA*, left coronary artery; *LV*, left ventricle; *Fwd*, forward; *Bwd*, backward; *Vmax*, maximum velocity; *TVI*, time-velocity integral; *Ved*, forward end-diastolic velocity.

94

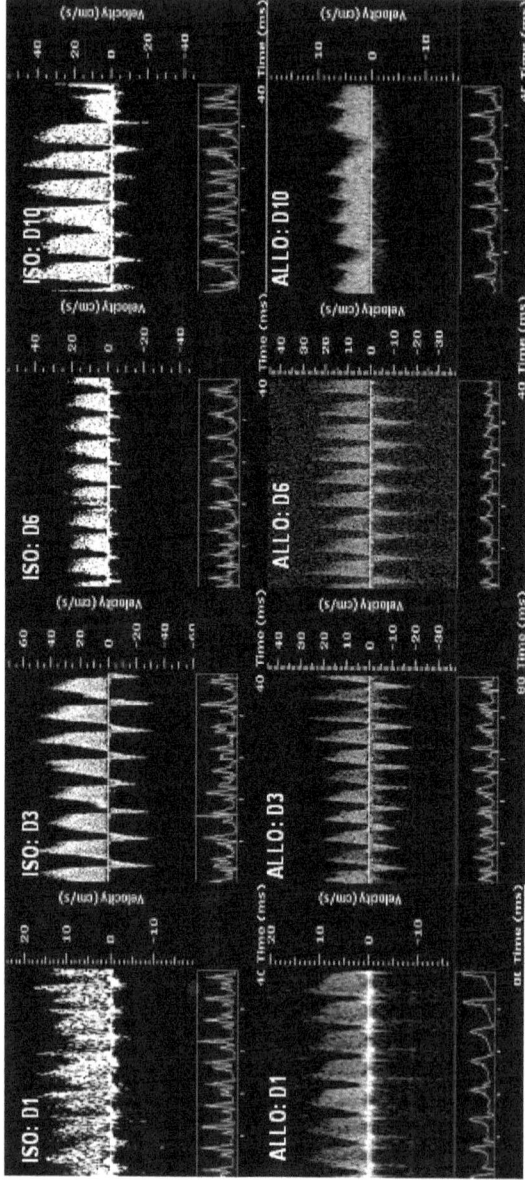

Figure 28 - Serial Doppler Characterization of Right Coronary Arterial flow in Isografts and Allografts. A. As with aortic and LCA flow, Doppler spectra produced by the asynchronous combination of native and graft heart rate produce periodic fluctuations. Also like LCA analyses, forward waveforms act to perfuse the myocardium whereas graft systole produces backward waveforms. Lack of backward flow in acutely rejecting grafts is also seen. Gaps in day 10 waveforms indicate a visibly identifiable periods of respiration. *ISO*, isografts; *ALLO*, allografts.

95

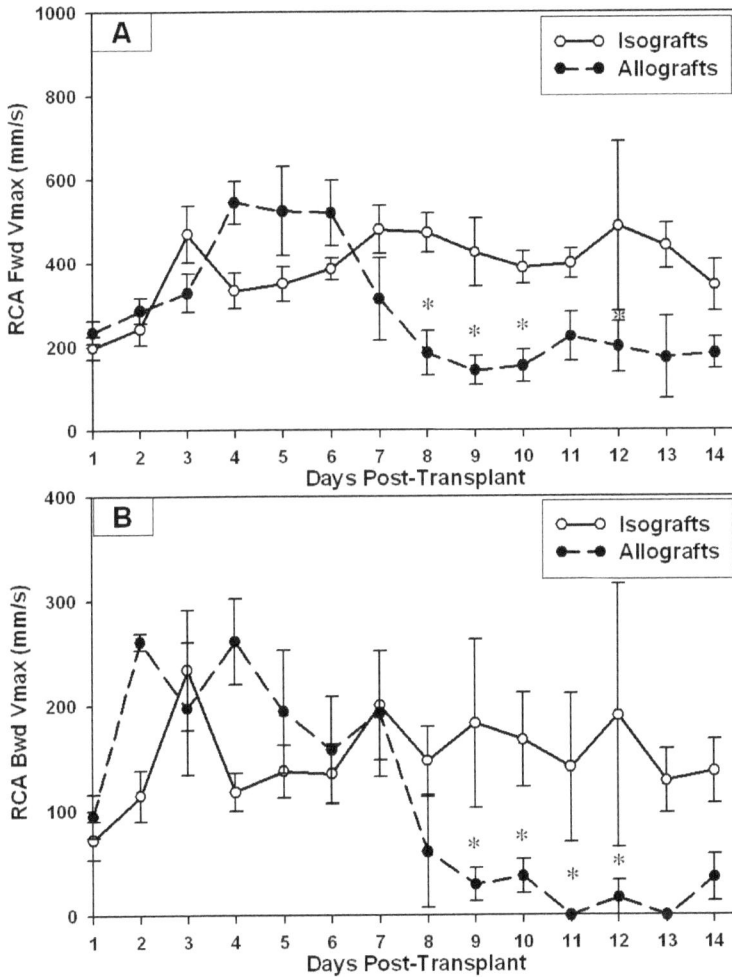

Figure 29 – Peak Right Coronary Artery Forward and Backward Doppler Velocities. A. Significant differences were detected on days 8-10 and 12 (*$P \leq 0.001$). **B.** Statistical differences in maximum backward velocity (Bwd Vmax) were detected from allografts on days 9-12 (*$P \leq 0.001$). Data are reported as means $\pm$ SEM.

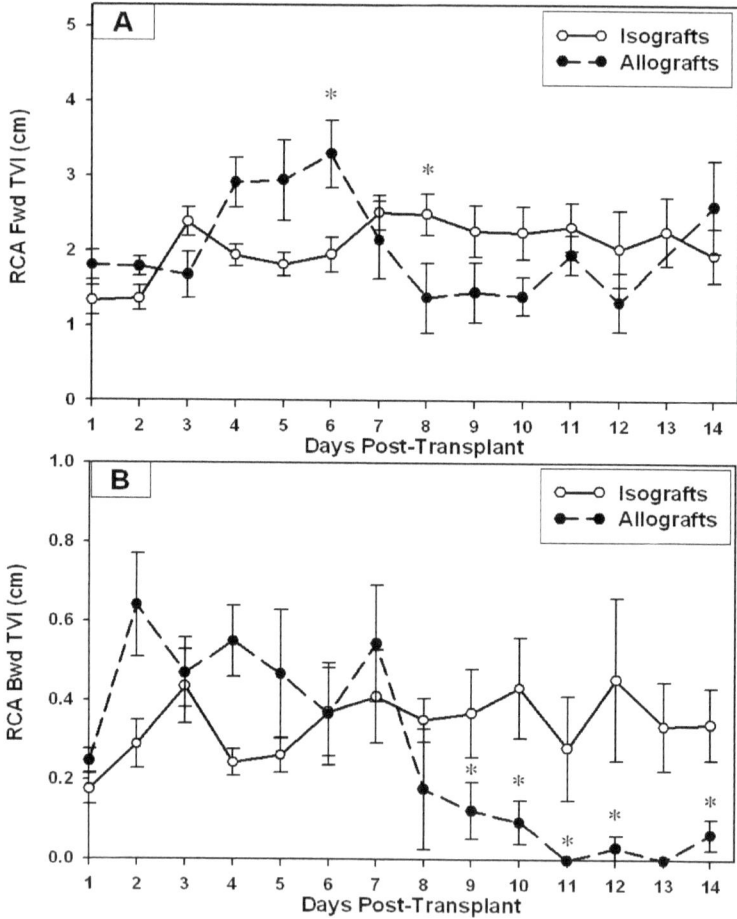

Figure 30 – Forward and Backward Time-Velocity Integrals (TVI) in the Right Coronary Artery. TVI traces were performed for maximum peak velocities (Vmax) of forward and backward waveforms. **A.** Only transient significant differences are seen on days 6 and 8 for Fwd TVI between iso- and allografts (*P=0.023). **B.** Allograft Bwd TVI significantly differs from isografts on day 9 until the end of the period of observation (*P=0.001). Data are reported as means ± SEM.

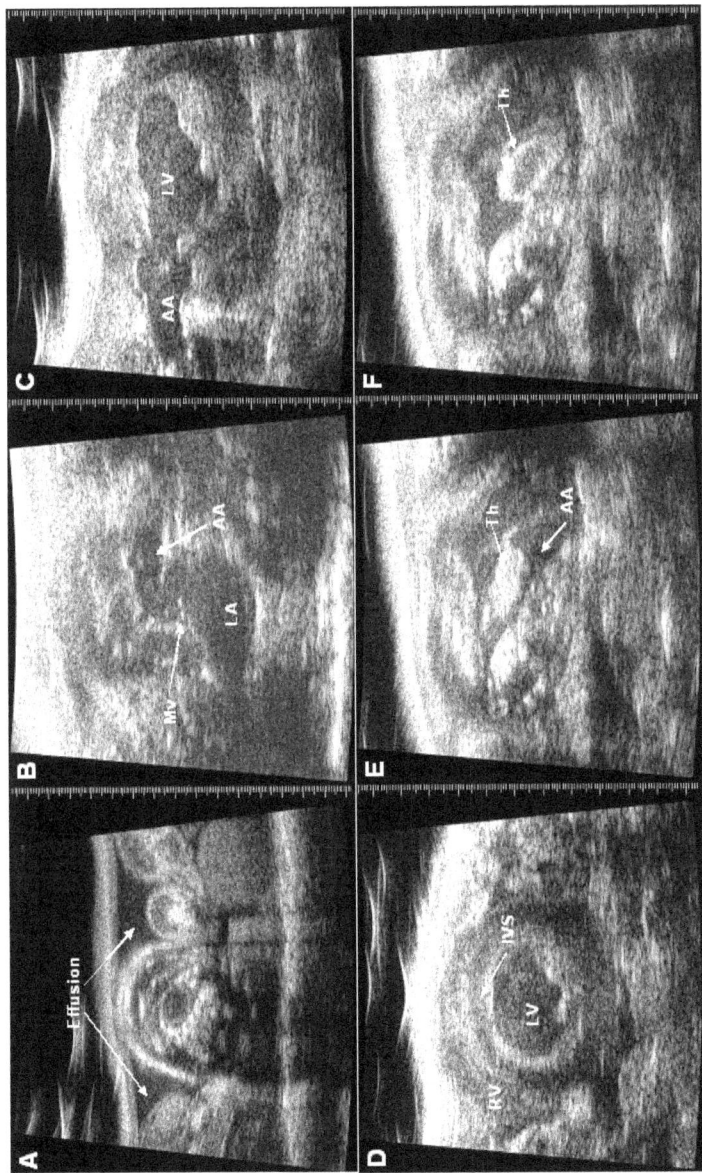

Figure 31 – Perigraft Effusion, Cardiac Dilatation, and Aortic Thrombi. A. Two-dimensional image of pericardial effusion taken from a day 6 allograft. **B.** Dilatation of the left atrial chamber in a day 13 isograft. **C, D.** Two-dimensional images in the long-axis (C) and short-axis views (D) showing left ventricular dilatation. The images were taken from an allograft 10 days post-transplant **E, F.** A dislodged ventricular thrombus seen traversing the aortic orifice and into the ascending aorta. The 2-D image was taken in a day 6 isograft during diastole (E) and systole (F). *Mv*, mitral valves; *LA*, left atrium; *AA*, ascending aorta; *LV*, left ventricle; *RV*, right ventricle; *IVS*, interventricular septum; *Th*, thrombus.

98

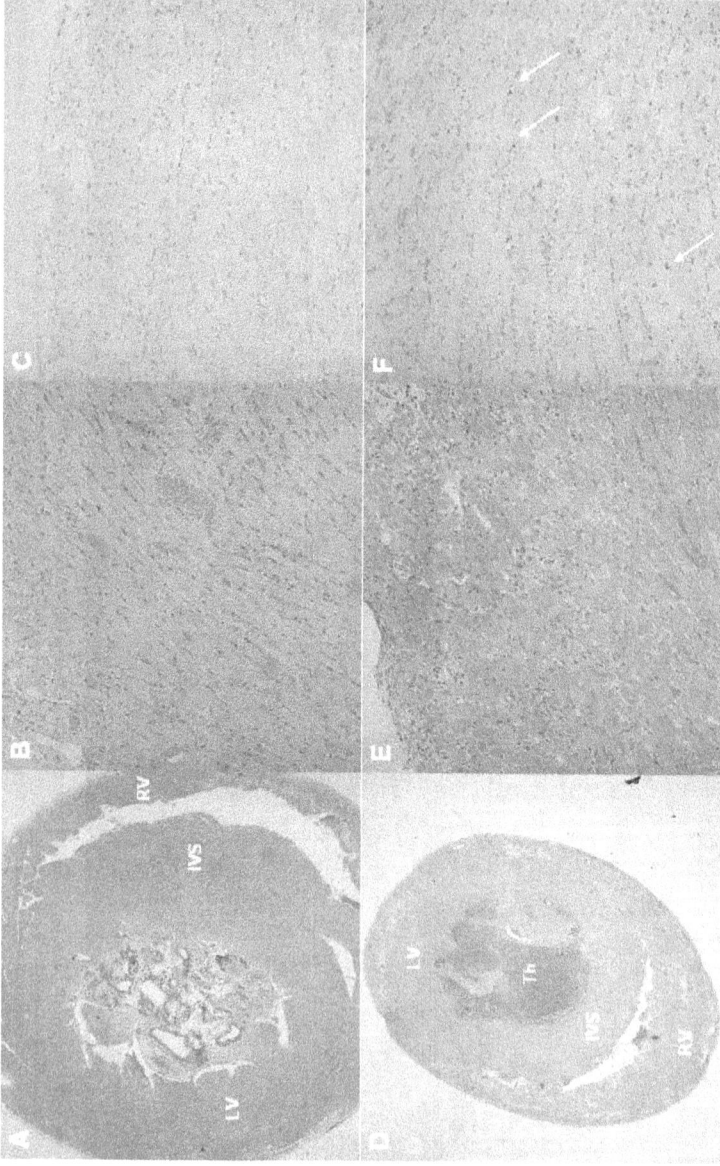

Figure 32 – H&E and TUNEL Staining of Murine Cardiac Grafts. Histology sections showing the short-axis of a day 14 cardiac isograft (**A**) and a day 9-rejected allograft (**D**). Magnification of H&E-stained sections of the same isograft (**B**) and allograft (**E**) tissue specimens. TUNEL-positive cells are indicated by white arrows in a day 6-rejected allograft (**F**), which are absent in the isograft specimen (**C**). *LV*, left ventricle; *RV*, right ventricle; *IVS*, interventricular septum; *Th*, thrombus.

99

Table VII. Summary Table of Haemodynamic, Morphologic and Physiologic Measurements

Parameter	Days Post-Transplant							
	1		3		6		10	
	ISO	ALLO	ISO	ALLO	ISO	ALLO	ISO	ALLO
General Measurements								
Number of viable grafts	8	10	8	10	7**	9	7**	6
Body weight (g)	21.1±0.9	21.9±1.1	20.4±0.7	21.0±0.9	20.7±0.6	21.4±0.8	21.4±1.0	23.4±0.9
Graft weight								
% Change			ISO: -13.3±8.2%		ALLO: 28.6±8.9%*			
Raw weight (g)			ISO: -0.023±0.014		ALLO: 0.0415±0.012*			
Native heart rate (BPM)	495±26	522±20	552±23	532±15	575±27	562±11	548±13	570±13
Graft heart rate (BPM)	421±17	410±19	541±29^a	567±22^a	562±20^a	482±10	554±33	407±36*^{b}
Palpation	4.0±0.0	4.0±0.0	2.9±0.2^a	3.6±0.12*	2.1±0.2^a	1.7±0.3^{a,b}	2.8±0.18^a	0.6±0.3*^{a,b,c}
Echogenicity (Mean arbitrary units)	1.15±0.18	1.04±0.16	2.25±0.16^a	1.76±0.17^a	2.33±0.18^a	2.01±0.16^a	2.18±0.2^a	2.27±0.20^a
Ascending Aorta								
Fwd Vmax (mm/s)	308±61	300±52	282±49	240±27	295±22	433±64	248±87	538±27*^{b}
Bwd Vmax (mm/s)	225±30	222±32	138±15	233±33	133±18	345±41*	110±13	310±81*
Fwd TVI (cm)	0.63±0.07	0.52±0.07	0.68±0.09	0.59±0.08	0.93±0.09	1.31±0.17	0.75±0.20	1.47±0.38*^{a,b}
Bwd TVI (cm)	0.50±0.05	0.41±0.07	0.25±0.02	0.38±0.05	0.27±0.04	0.56±0.09	0.21±0.03	0.92±0.23*^{a,b}
Fwd RI	0.92±0.03	0.86±0.04	0.87±0.02	0.86±0.03	0.82±0.02	0.83±0.02	0.79±0.04	0.92±0.03*
Fwd Ved (mm/s)	23±8	40±12	32±3	38±12	51±5	67±6	44±11	61±24
Left Coronary Artery (LCA)								
Fwd Vmax (mm/s)	178±30	235±22	383±70	484±34	599±67^a	602±96^a	432±77	251±62*
Bwd Vmax (mm/s)	51±9	43±12	215±63	202±41	216±25	281±51^a	190±49	71±35^c
Fwd TVI (cm)	1.37±0.29	2.00±0.25	1.59±0.22	2.45±0.10	2.70±0.22	3.04±0.48	2.57±0.60	2.32±0.65
Bwd TVI (cm)	0.14±0.02	0.17±0.06	0.47±0.14	0.44±0.07	0.53±0.07	0.62±0.12	0.44±0.14	0.23±0.09
Fwd RI	0.48±0.02	0.55±0.03	0.52±0.03	0.57±0.03	0.47±0.04	0.61±0.03*	0.55±0.02	0.53±0.04
Fwd Ved (mm/s)	93±19	107±16	187±42	209±24	306±26^a	239±41	230±47	121±35
Right Coronary Artery (RCA)								
Fwd Vmax (mm/s)	198±28	235±27	470±68	329±46	387±26	520±78	389±39	154±39*
Bwd Vmax (mm/s)	68±17	95±21	234±58	197±63	135±28	158±51	167±45	37±16*
Fwd TVI (cm)	1.33±0.20	1.81±0.20	2.39±0.19	1.68±0.32	1.96±0.23	3.30±0.45*	2.25±0.35	1.40±0.25
Bwd TVI (cm)	0.18±0.03	0.25±0.03	0.44±0.09	0.47±0.09	0.37±0.11	0.37±0.13	0.43±0.13	0.096±0.05*
Fwd RI	0.61±0.04	0.52±0.04	0.48±0.03	0.46±0.03	0.50±0.02	0.49±0.02	0.51±0.04	0.49±0.03
Fwd Ved (mm/s)	80±14	109±9	254±45^a	183±30	194±18	261±35	196±31	61±12*
Mitral Valve								
MR (mm/s)	759±158	909±405	1655±246	1265±300	994±345	1249±402	2816±565	171±171*
Aortic Valve								
AR (mm/s)	2710±60	2133±381	2312±419	2265±313	1988±474	3430±258	2435±522	1890±633
Main Pulmonary Artery (MPA)								
Vmax (mm/s)	223±65	241±76	414±74	503±151	773±212^a	546±139	343±82	302±130
TVI (cm)	0.70±0.17	1.04±0.29	1.85±0.28	1.85±0.36	3.10±0.55^a	2.14±0.66	1.45±0.32	1.47±0.76
Tricuspid Valve								
E-wave Vmax (mm/s)	35±8	43±5	50±2	59±8	76±12	101±22^a	51±5	33±33^c
A-wave Vmax (mm/s)	109±17	122±12	125±23	168±30	213±32	163±45	166±41	72±72
E/A ratio	0.39±0.15	0.37±0.05	0.46±0.10	0.41±0.10	0.42±0.10	0.69±0.09	0.51±0.21	0.46±0.00^c
TVI (cm)	0.31±0.06	0.39±0.04	0.39±0.07	0.49±0.08	0.53±0.06	0.58±0.13	0.48±0.07	0.43±0.43
M-mode LV Measurements								
PWes (mm)	1.20±0.05	1.49±0.08	1.23±0.04	1.69±0.07*	1.36±0.11	2.03±0.06*^{a}	1.11±0.09	1.85±0.15*
PWed (mm)	1.14±0.04	1.41±0.09	1.21±0.05	1.56±0.07*	1.27±0.09	1.94±0.07*^{a,b}	1.02±0.09	1.73±0.13*
AWes (mm)	0.97±0.07	1.20±0.05	1.08±0.07	1.49±0.08*	1.06±0.07	1.66±0.12*^{a}	1.01±0.12	1.43±0.20*
AWed (mm)	0.91±0.04	1.13±0.06	0.99±0.06	1.34±0.08*	1.00±0.06	1.54±0.10*^{a}	0.92±0.11	1.36±0.22*
ESD (mm)	2.07±0.18	1.55±0.12	1.23±0.13	1.27±0.23	1.28±0.22	1.01±0.20	1.26±0.19	1.89±0.25*^{c}
EDD (mm)	2.31±0.18	1.78±0.11	1.50±0.09	1.58±0.25	1.51±0.20^a	1.31±0.17	1.52±0.17	2.17±0.23*^{a,b,c}
SV (μL)	5.04±1.18	2.89±0.42	2.22±0.62	3.76±1.92	2.24±0.74	1.85±0.32	2.38±0.50	4.35±0.85
CO (mL/min)	2.61±0.60	1.46±0.21	1.25±0.39	2.06±0.94	1.45±0.46	1.04±0.22	1.41±0.32	2.04±0.32
%EF	24.2±4.0	28.7±3.7	36.3±9.0	44.8±9.3	37.1±9.6	49.9±8.8	39.9±7.5	30.8±7.1
%FS	10.8±2.0	12.2±1.6	20.3±5.7	20.3±5.7	18.8±6.2	26.2±6.1	18.7±4.3	14.0±3.6

Abbreviations: ISO, isografts; ALLO, allografts; Fwd, forward; Bwd, backward; Vmax, maximal velocity at peak systole; Ved, end-diastolic velocity; RI, resistance index;
TVI, time-velocity integral; MR, mitral regurgitation; AR, aortic regurgitation; PWes, end-systolic left ventricular posterior wall thickness; PWed, end-diastolic posterior wall
thickness; AWes, end-systolic left ventricular anterior wall thickness; AWed, end-diastolic left ventricular anterior wall thickness; ESD, left ventricular end-systolic diameter;
EDD, left ventricular end-diastolic diameter; SV, stroke volume; CO, cardiac output; %EF, %ejection fraction; %FS, %fractional shortening.

*indicates significant difference versus isografts ($P<0.05$).

**one isograft died for reasons unrelated to surgery

a, b, c, and d represent significant differences ($P<0.05$) vs. days 1, 3, 6, and 10 post-transplant, respectively.

Data are reported as mean ± SEM.

100

Table VIII. Summary Statistics and Interaction Terms

Parameter	Time			Treatment			Time x Treatment		
	F	P	Significance	F	P	Significance	F	P	Significance
General Measurements									
Body Weight (g)	4.822	<0.001	Yes	0.388	0.542	No	0.381	0.974	No
Graft Weight									
% Change	--	--	--	t=3.368	0.0098	Yes	--	--	--
Raw Weight (g)	--	--	--	t=3.356	0.0100	Yes	--	--	--
Native heart rate (BPM)	2.109	0.017	Yes	0.063	0.805	No	0.322	0.988	No
Graft heart rate (BPM)	6.93	<0.001	Yes	19.594	<0.001	Yes	6.07	<0.001	Yes
Palpation	46.106	<0.001	Yes	38.211	<0.001	Yes	21.889	<0.001	Yes
Echogenicity	18.117	<0.001	Yes	1.049	0.32	No	0.964	0.42	No
Ascending Aorta									
Fwd Vmax (mm/s)	2.239	0.011	Yes	11.887	0.003	Yes	3.4	<0.001	Yes
Bwd Vmax (mm/s)	0.91	0.544	No	26.625	<0.001	Yes	1.467	0.138	No
Fwd TVI (cm)	3.088	<0.001	Yes	10.609	0.004	Yes	2.252	0.011	Yes
Bwd TVI (cm)	2.289	0.009	Yes	39.727	<0.001	Yes	3.709	<0.001	Yes
Fwd RI	0.91	0.545	No	2.054	0.169	No	1.59	0.096	No
Fwd Ved (mm/s)	0.827	0.631	No	4.136	0.054	No	0.173	0.999	No
Left Coronary Artery (LCA)									
Fwd Vmax (mm/s)	6.505	<0.001	Yes	0.635	0.436	No	2.223	0.011	Yes
Bwd Vmax (mm/s)	6.772	<0.001	Yes	3.5	0.078	No	2.373	0.007	Yes
Fwd TVI (cm)	3.156	<0.001	Yes	3.036	0.097	No	1.346	0.195	No
Bwd TVI (cm)	5.413	<0.001	Yes	2.103	0.163	No	1.763	0.055	No
Fwd RI	0.645	0.812	No	4.021	0.057	No	1.78	0.052	No
Fwd Ved (mm/s)	5.489	<0.001	Yes	2.982	0.1	No	1.749	0.058	No
Right Coronary Artery (RCA)									
Fwd Vmax (mm/s)	3.22	<0.001	Yes	1.302	0.26	No	3.554	<0.001	Yes
Bwd Vmax (mm/s)	3.193	<0.001	Yes	1.789	0.188	No	2.979	0.001	Yes
Fwd TVI (cm)	2.09	0.023	Yes	0.12	0.731	No	3.132	<0.001	Yes
Bwd TVI (cm)	2.927	0.001	Yes	1.755	0.192	No	3.499	<0.001	Yes
Fwd RI	1.291	0.245	No	0.0854	0.773	No	1.082	0.383	No
Fwd Ved (mm/s)	4.813	<0.001	Yes	1.305	0.263	No	4.497	<0.001	Yes
Mitral Valve									
MR (mm/s)	1.456	0.154	No	22.227	<0.001	Yes	3.747	<0.001	Yes
Aortic Valve									
AR (mm/s)	2.028	0.023	Yes	1.1	0.309	No	2.847	0.001	Yes
Main Pulmonary Artery (MPA)									
Vmax (mm/s)	2.679	0.004	Yes	5.12	0.041	Yes	1.112	0.364	No
TVI (cm)	2.773	0.003	Yes	6.839	0.019	Yes	1.243	0.269	No
Tricuspid Valve									
E-wave Vmax (mm/s)	3.963	<0.001	Yes	2.452	0.131	No	2.169	0.028	Yes
A-wave Vmax (mm/s)	1.84	0.053	No	3.674	0.082	No	1.461	0.155	No
E/A ratio	2.841	0.005	Yes	1.079	0.315	No	1.728	0.087	No
TVI (cm)	1.574	0.114	No	0.388	0.546	No	1.386	0.188	No
M-mode LV Measurements									
PWes (mm)	6.006	<0.001	Yes	73.052	<0.001	Yes	1.593	0.094	No
PWed (mm)	6.632	<0.001	Yes	78.048	<0.001	Yes	1.937	0.031	Yes
AWes (mm)	4.603	<0.001	Yes	21.124	<0.001	Yes	1.48	0.132	No
AWed (mm)	4.143	<0.001	Yes	21.165	<0.001	Yes	1.295	0.222	No
ESD (mm)	4.542	<0.001	Yes	3.532	0.077	No	4.369	<0.001	Yes
EDD (mm)	5.463	<0.001	Yes	4.179	0.057	No	5.033	<0.001	Yes
SV (μL)	2.95	<0.001	Yes	1.451	0.245	No	2.742	0.002	Yes
CO (mL/min)	4.047	<0.001	Yes	1.406	0.252	No	3.41	<0.001	Yes
%EF	1.35	0.191	No	0.189	0.669	No	1.367	0.182	No
%FS	1.51	0.12	No	0.178	0.679	No	1.291	0.225	No

Abbreviations: ISO, isografts; *ALLO,* allografts; *Fwd,* forward; *Bwd,* backward; *Vmax ,* maximal velocity at peak systole; *Ved,* end-diastolic velocity; RI, resistance index; *TVI,* time-velocity integral; *MR,* mitral regurgitation; *AR,* aortic regurgitation; *PWes,* end-systolic left ventricular posterior wall thickness; *PWed ,* end-diastolic posterior wall thickness; *AWes,* end-systolic left ventricular anterior wall thickness; *AWed,* end-diastolic left ventricular anterior wall thickness; *ESD ,* left ventricular end-systolic diameter; *EDD,* left ventricular end-diastolic daimeter; *SV,* stroke volume; *CO ,* cardiac output; *%EF,* %ejection fraction; *%FS ,* %fractional shortening.

IV. DISCUSSION & CONCLUSION

4.1 Novelty and Relevance of Study

This is the first report to show the utility of a state-of-the-art high-frequency UBM imaging modality to characterize serially and non-invasively acute cardiac allograft rejection in a common model of heterotopic transplantation. We hypothesized that echocardiographic parameters procured by UBM and related functions could detect differences between accepted isografts and acutely rejecting allografts not only earlier, but can also yield substantially more important information on graft viability than the common method of finger palpation. Our main objective was to utilize UBM modality to provide comprehensive information on the dynamic changes that occur with respect to haemodynamic flow, morphology and dimensions, and general graft physiology in murine cardiac grafts during immune acceptance of isografts and allograft rejection. Furthermore, it is our intention that this information will enable the establishment of a timeline of echocardiographic changes to aid investigators in interpreting data from studies on heterotopic murine heart transplant models. Taken together, the acquisition of reliable parameters for assessing cardiac graft rejection and a systematic time-course of changes in graft function would allow for a more critical and accurate determination of the affects of genetic and molecular manipulation as well as pharmacological treatments on graft survival in this model.

4.2 Assessing Graft Viability by Palpation and UBM-Doppler

Serial measurements of finger palpation scores indicate that isografts and allografts cannot be differentiated until day 7 PTx using this method (**Fig. 3**). Generally, our palpation values are in agreement with earlier studies (Scherrer-Crosbie *et al.,* 2002). We also found that for allografts, the median survival time obtained by palpation was 2 days shorter than the median survival time of 12 days as determined by UBM-Doppler, ECG and B-mode imaging, though this difference was not significant (**Fig. 4, Table V**). It should be noted that the echo-derived median survival time of 12 days has been previously reported for the same recipient-donor strain combination using finger palpation (Alexander *et al.,* 1996; Pearson *et al.,* 1994), suggesting that UBM-Doppler does not confer a time advantage in detecting graft death over the traditional method. However, this finding should not be assumed to indicate

102

that UBM imaging is not useful. Our premise was that ultrasound would enable detailed assessment of graft function otherwise not possible by palpation, histology and other methods, and that it could detect subtle differences earlier in the PTx time course. Furthermore, studies on the state of canine and rodent HHTx allografts early post-transplant indicate depressed systolic and diastolic function characterizes these grafts (Alyono *et al.,* 1984; Scherrer-Crosbie *et al.,* 2002; Zhou *et al.,* 2007; Szabo *et al.,* 2001), and yet palpation scores are highest during the same time period, suggesting that data gained from finger palpitations may reflect large observer bias (*i.e.,* expectation of good contractility prior to the onset of rejection) and is thus an inaccurate method of determining graft viability. In any case, it would seem that palpation scores can be used for general assessment of graft contractility as both isografts and allografts showed decreasing scores after day 1 PTx. This would concur with the expected effects of ischemia-reperfusion (I/R)-injury and formation of LV cavity thrombus (see below) (Smith *et al.,* 1987).

4.3   Heart Rate Comparisons

In order to validate which UBM-Doppler waveforms were products of graft or native cardiac cycles, we needed to determine if the ECG signals from graft and recipient hearts could be differentiated using continuous, real-time analysis of ECG signals from different locations in the same recipient. We carried out this short experiment in anaesthetized animals using needle electrodes inserted subcutaneously in the lower abdomen and limbs (Abbott *et al.,* 1965; Superina *et al.,* 1986) and via intracardiac leads. The instantaneous abdominal-intracardiac ECG (**Fig. 5A**) and limb ECG (**Fig. 5B**) performed on isografts and allografts verified two aspects of the study: first, a physiologic acquisition system can detect the voltage complexes due to recipient and donor hearts as unique, separate waveforms and second, limb/paw ECG, as employed in the UBM system, included these two waveforms on the same ECG spectrum. This offered us the opportunity to study these heart signals with the purpose of determining which components of Doppler waveforms and time-course M-modes arise from the separate and unique combination of graft and native cardiac cycles (Zhou *et al.,* 2007). Analysis of ECG also provided some information on the manifestation of cardiac abnormalities, such as bradycardia and arrhythmia (**Fig. 5C**). Cardiac arrhythmias observed in both experimental groups several hours after transplantation have been reported in rat

(Heron, 1972) and murine HHTx recipients (Superina *et al.*, 1986), though this was not specifically found in our experiment. Superina *et al.* also documented the appearance of regular ECG waveforms usually seen after the immediate post-operative period for isografts and allografts. Later in the PTx period, Mottram *et al.* observed that murine HHTx allografts underwent a rapid fall in palpated HR, measured HR, and ventricular complex voltage in 10-14 days PTx due to acute rejection (Mottram *et al.*, 1988); this finding is in contrast with our data, which showed declining measured allograft HR from day 7 onwards. It is possible that UBM-Doppler-derived measurements of graft HR are more sensitive to changes in graft viability than graft HR obtained by ECG electrodes, indicating that morphologic and haemodynamic changes within the graft may precede the development of sinus node dysfunction as a result of acute cellular rejection.

The significant rise from lower (~400 BPM) to higher (~550BPM) graft HR seen in both groups of grafts from days 1-3 (**Fig. 6B**) probably indicated recovery from perioperative ischemia. This finding corroborates results from our previous study, which showed lower isograft HR on day 1 versus day 5, 15 and 50 (Zhou *et al.*, 2007) as well as work done by Alyono *et al.*, where they previously reported depressed diastolic and systolic function in dog HHTx grafts immediately following transplantation (Alyono *et al.*, 1984). The authors attributed the change to trauma from operative ischemia rather than immune rejection and further reported gradual amelioration of graft function and HR to levels near that achieved by control hearts after three days post-implantation (Alyono *et al.*, 1984), a similar time frame of recovery also suggested by the peak HR achieved on day 3 in both study groups in our experiment. Interestingly, data from rat HHTx models showed that the low-frequency/high-frequency power ratio and % low-frequency/total power ratios – quantities reflecting the spectral power of RR intervals of the ECG spectrum and computed using a fast-Fourier transform algorithm – were similar for both allografts and isograft controls on day 3, and that these parameters subsequently increased significantly for allografts in comparison to isografts as acute rejection became more prominent (Wada *et al.*, 1999).

With the exception of the first 3 days PTx, graft HR in isografts was consistently similar to native HR figures (~500-600BPM) in anesthetized recipients. Because cardiac grafts are denervated, they lack the inhibitory input from autonomic vagal nerve stimulation and are less affected by anaesthesia, thus yielding a high graft basal HR that is comparable to isoflurane-anaesthetized HR from normal mouse hearts (Hunt, 2001; Roth *et al.*, 2002). Expectedly, native HR for isograft or allograft recipients did not change with respect to the experimental groups (**Fig. 6A**).

4.4   Changes in Myocardial Mass in Allografts and Isografts

Historically, increased graft and LV masses due to inflammatory cell infiltration and myocardial edema were often associated with clinical rejection (Sagar *et al.*, 1981; Burgess *et al.*, 2002). By using a digital scale to weigh the small mouse hearts, we observed that allograft raw and % weight changes were significantly higher compared to isografts and furthermore, allografts underwent significant weight increases compared to measurements taken before implantation (**Fig. 7**). It was important to note only changes in, rather than absolute values of, graft mass since the hearts originate from different donor strains. Augmented myocardial mass in rejecting allografts has also been shown in studies involving rat HHTx grafts, where myocardial water content was significantly higher in allografts compared to isografts after 5 days PTx (Szabo *et al.*, 2001). The authors also found that rat allografts have histologically-confirmed mild to moderate rejection after 3 days and severe acute rejection after 5 days PTx, which led the authors to propose a typical sequence of events in rat allograft acute rejection: recovery from ischemia; generation of edema; mild to moderate rejection at day 3 PTx with associated impaired diastolic compliance (*i.e.*, decreased relaxation time constant of the LV pressure fall); and finally severe systolic dysfunction at day 5 PTx (*i.e.*, decreased slope of the systolic pressure–volume relation) (Szabo *et al.*, 2001). The cause for decreased diastolic compliance seen in acutely rejecting grafts was noted in a separate study to be caused likely by changes in the elastic properties of the myocardium during ACR, with generation of myocardial edema, exudates, and cellular infiltration at least partly responsible for this change (Amirhamzeh *et al.*, 1994). Our results indicated that the change in isograft weight (pre-implantation mass minus post-explantation mass) was not significant. Work by Korecky *et al.*, however, have shown that myocardial atrophy, which was observed one week after transplantation of rat HHTx isografts, led to a significant decrease in LV mass - 59% of the weight of control *in situ* hearts (Korecky and Masika, 1991). It should be noted that the current study's small sample size (n=5, isografts) and the inherent variability with cutting at the exact line of the anastomoses may have rendered the data incapable of determining a significant decrease in isograft weight.

Analyses carried out at the cellular level illustrated the possible mechanisms for the decreased isograft weights and the remarkable plasticity of the cardiac graft. For example, Rakusan *et al.* have determined that cellular changes in atrophied cardiac isografts may include increased capillary and myocyte density, increased numbers of nucleated myocytes located close to the nearest capillary, as well as an involution of coronary capillaries (Rakusan *et al.*, 1997). It was previously thought that rodent grafts were incapable of such large increases and decreases in cardiac mass within a small time frame, thus substantiating the need to evaluate critically the physical and haemodynamic status of these grafts in studies that manipulate morphological and physiologic properties of cardiac grafts (*e.g.*, pressure and volume-induced cardiac hypertrophy).

4.5   *Increased Cardiac Backscatter Following Heterotopic Transplantation*

To our knowledge, this is the first report of increased echo backscatter in live, real-time imaging of an experimental rodent solid organ graft. Clinically, increased end-diastolic posterior wall and septal 2D-integrated backscatter has been used to identify mild, moderate and severe acute rejection though this approach has not yet been validated for diagnosing rejection (Angermann *et al.*, 1997). In our experimental model, we observed that semi-qualitative mean scores of myocardial intensity sampled from the mid-segment of the anterior wall of the LV significantly increased after day 1 for both study groups (**Fig. 5**). The observation of immediate increases in backscatter following transplantation was also reported in HHTx goat models (Bergese *et al.*, 1997; Hete *et al.*, 1990). Previously, studies have suggested that apoptosis, and the ensuing subcellular nuclear changes, condensation, and cellular fragmentation (Czarnota *et al.*, 1997; Czarnota *et al.*, 1999), likely causes an increase in ultrasound backscatter signal. We verified the presence of apoptotic cells in acutely rejected allografts (median survival time = 12 days PTx) (**Fig. 32F**), which were absent in 14 day PTx isografts (**Fig. 32C**). Apoptosis of cardiomyocytes and endothelial cells due to surgery-induced I/R injury (Scarabelli *et al.*, 2002) and cytotoxic T cell-mediated destruction of graft tissues (Bergese *et al.*, 1997) have been documented, though resorption of I/R-induced apoptotic cells is likely the reason why no TUNEL positive cells were found in two week PTx isografts (White *et al.*, 1997; Gottlieb *et al.*, 1994).

The role of apoptosis in the progression of acute rejection is unclear (Bergese *et al.*, 1997; Tanaka *et al.*, 2005c; Yamaura *et al.*, 2004; Raisky *et al.*, 2004). Bergese *et al.*, for example, have noted that although apoptosis may not be indicative of graft rejection, periarterial clustering of apoptotic cells in accepted grafts may reflect immunoregulatory processes that maintain graft acceptance or repair processes that promote chronic vascular remodelling (Bergese *et al.*, 1997). It is very likely that due to the persistently high scores in echo intensity in accepted isografts post-transplant in our experimental scheme, cellular processes that change the acoustic properties of graft tissues are not exclusive to acutely rejecting grafts and apoptosis is not a sufficient explanation for the changes in backscatter seen in the PTx period. Alternate and/or contributory mechanisms such as necrosis have been shown to increase ultrasound backscatter signal in heat-killed human acute myeloid leukaemia cells (Czarnota *et al.*, 1997). In acutely rejecting murine HHTx grafts, infiltrating polymorphonuclear cells were associated with necrosis of tissue parenchyma due to ischemia, and may also represent an important mechanism for the apparent increased echo scores, though we did not verify this mechanism in our study (Miura *et al.*, 2003). Other observations, such as myofibrillar changes (O'Brien *et al.*, 1995) and myocardial edema (Chandrasekaran *et al.*, 1987) have also been associated with increased backscatter in rejecting grafts.

4.6 Post-Transplant Changes in LV Wall Thicknesses and Chamber Dimensions

4.6.1 Ischemia-Reperfusion Injury and Early Morphologic Changes

Despite the exceptional high-resolution of UBM, early and long-term histologic changes of the graft tissue often precluded delineation of epicardial and endocardial borders. Image quality generally became less clear as time progressed, especially in acutely rejecting allografts. The interobserver variability was determined to be 0.129 ± 0.284 mm and the intraobserver variability 0.06 ± 0.151 mm, which is higher than previous studies on murine HHTx grafts (Scherrer-Crosbie *et al.*, 2002). Increases in anterior wall thicknesses at end-systole (AWes) and end-diastole (AWed) were observed from days 3 to 10 and 14 PTx (**Fig. 12A, B**) and posterior wall thicknesses (PWes, PWed) from day 2 PTx onwards in allografts in comparison to isografts (**Fig. 13A, B**). The general observation of a similar pattern of changes between end-systolic and end-diastolic wall thicknesses for each respective ventricular wall may reflect myocardial stiffening and fibrosis present in these heterotopic grafts (Smith *et al.*, 1987). Furthermore, both walls of the LV from both experimental groups showed a general biphasic pattern: an initial increase in wall thicknesses to a maximum value midway in the PTx period (range 5-7 days); and a later decrease to values comparable to early PTx figures. Several studies have noted that increased LV wall thickness of cardiac grafts following heterotopic transplantation is very likely attributable to the effects of I/R injury and consequent vascular permeability and interstitial edema (Scherrer-Crosbie *et al.*, 2002; Tanaka *et al.*, 2005a; Metzler *et al.*, 2001; Burgess *et al.*, 2002). This is also confirmed by the timeline suggested by Szabo *et al.* in rat HHTx allografts (Szabo *et al.*, 2001) and is demonstrated by our observation that both isografts and rejecting allografts underwent initial increases in wall thicknesses – suggesting that the initial changes are not mediated by ACR. Admittedly, the early increases in wall thicknesses in isografts did not reach significance. Interestingly however, a previous report by Smith *et al.* noted that histopathological damage induced by I/R injury in control isografts was observed not only early PTx, but could also be detected as late as one month after transplantation (Smith *et al.*, 1987). In this study, Smith *et al.* showed that when heterotopically transplanted murine isografts had ischemic times of 30 and 60 minutes, histological examination showed minimal myocardial damage, with necrosis or scarring occupying less than 5% of the cross-sectional area of hearts bisected from the apex to base at day 7 PTx. With 120 minutes of ischemia, however, maximal ischemic damage was also observed at 7 days PTx, but with more than 30% myocardial necrosis or scarring of myocardial tissue. Regardless of ischemic times, all isografts showed a fine, generalized, perimyocytic fibrosis (Smith *et al.*, 1987). The experimental scheme employed in our study utilized about 30-45 minutes ischemic time, which would suggest that I/R-induced myocardial necrosis or cell death was scarce in isografts, as confirmed by TUNEL assays (**Fig. 32C**) - a common laboratory technique which has been purported to lack the ability for discriminating apoptotic from necrotic cells (Grasl-Kraupp *et al.*, 1995).

It is important to note that I/R injury can have a substantial effect on the status of the graft. At the cellular and molecular level, I/R injury generates a highly dynamic environment which gives rise to the expression of inflammatory mediators, such as chemokines and cytokines (*e.g.,* tumour necrosis factor-alpha, interleukin-1β, and monocyte/macrophage chemoattractant protein-1) (Tanaka *et al.,* 2005a), an immediate elevation in serum troponin T, creatine kinase and lactate dehydrogenase isoenzyme 1 levels (Metzler *et al.,* 2001) and a loss of high-energy phosphates (creatine phosphate, adenosine triphosphate) (Galinanes and Hearse, 1991). Later changes in the metabolic profiles of post-ischemic grafts are usually accompanied by a return to pre-ischemic values and a subsequent decline over the PTx period in isografts (Galinanes and Hearse, 1991). It should be noted that the these metabolic changes are critical for mediating I/R injury; studies illustrating the effects of the treatment with selective protein kinase C regulators, for example, have been shown to ameliorate post-ischemic damage induced by the loss of energy phosphates (Tanaka *et al.,* 2005a).

With respect to changes in ventricular dimensions early PTx, initial reductions of LV dimensions in hemodynamically-unloaded grafts have been documented (Spencer *et al.,* 2003; Welsh *et al.,* 2001b; Zhou *et al.,* 2007; Scherrer-Crosbie *et al.,* 2002). A similar pattern of change for LV dimensions (**Fig. 14A, B**) was obtained in our study, although it did not reach statistical significance. The general early trend for decreased ESD and EDD in both groups of grafts may be explained by the loss of loading conditions and/or the possible interference of the ventricular thrombus with systolic contraction and diastolic relaxation (see **Fig. 9B, D**) (Zhou *et al.,* 2007).

4.6.2   Late Changes are due to ACR and Ventricular Atrophy

The continued increases in wall thicknesses in allografts subsequent to the first several days PTx (i.e., after day 3) are very likely due to the effects of histopathological changes brought upon by ACR - infiltration of host lymphocytes and macrophages as well as myocyte necrosis, apoptosis, haemorrhage and cellular swelling (Asante-Korang, 2004; Del Castillo *et al.,* 1989; Scherrer-Crosbie *et al.,* 2002). With the exception of the present study and a previous study by our laboratories (Zhou *et al.,* 2007), Scherrer-Crosbie *et al.* reported the first and only study known to date to utilize echocardiographic imaging to investigate the potential of ultrasound in detecting changes in LV wall thicknesses in murine HHTx grafts during acute rejection. Using serial echocardiography, the authors detected a significant increase in LV posterior wall thickness in allografts versus isografts as early as day 1 PTx as well as a 20% and 65% increase from baseline in isografts and allografts, respectively, when

109

compared to day 5 PTx figures (Scherrer-Crosbie *et al.*, 2002). However, the poor resolution of the ultrasound system they used precluded reliable measurements from distorted images apparent in acutely rejecting day 5 allografts. Consequently, the authors implied that the thickness of the posterior wall in murine allografts continued to increase past day 6 PTx as acute rejection became more pronounced in allografts (Scherrer-Crosbie *et al.*, 2002). This finding contradicts our observation - that LV wall thicknesses actually undergo a biphasic pattern of initial increase and subsequent decrease - which stands as a unique finding previously undocumented in a rodent model of heterotopic cardiac transplantation. We suggest that, whereas immune-mediated increases in LV wall thicknesses in allografts predominate in the time period following the post-ischemic phase, chronic ventricular atrophy is a likely cause of later decreases in myocardial wall thicknesses. Previous studies suggest that the long-term decreases in LV mass of isografts is due to chronic histological changes causing ventricular atrophy as a result of under-filling and absence of physiologic ventricular pressure (Klein *et al.*, 1991; Rakusan *et al.*, 1997), though developed pressure and the maximum rate of pressure development were unaffected in atrophic ventricles as assessed by a pressure transducer connected to a polyethylene cannular-needle system inserted into the ventricular cavity of rat HHTx grafts (Kolar *et al.*, 1993).

Earlier experiments have validated the association between ventricular atrophy and haemodynamic load in a transplant model. For example, Korecky *et al.* have been successful in reducing the degree of atrophied ventricular myocardium by using an inserted polyethylene cannula into the graft aortic orifice to induce valvular incompetence and/or aortic stenosis. Left ventricular systolic pressure, rate-pressure product, and a greater retention of LV mass (85% of *in situ* hearts) was achieved in comparison to control isografts, indicating that the increased load significantly decreased LV atrophy and that this effect is related to increased ventricular load as suggested by higher peak LV pressures and higher rate-pressure products (Korecky and Masika, 1991). Other studies have also made similar associations between haemodynamic load and ventricular chamber dimensions, compliance, and mass, when aortic regurgitation was artificially induced by puncturing the aortic valves prior to implantation (Spencer *et al.*, 2003). It should be noted that one could postulate that the presence of high AR jet velocities (>2000 mm/s) from days 1-9 post transplant could have been, in principle, responsible for inducing LV hypertrophy in allografts in our study

110

(**Fig. 13A, B**) as opposed to immunologically-mediated events. This explanation, however, is problematic primarily because of two reasons: isografts, which also displayed chronically high AR, did not undergo increases in wall thicknesses and second, the large ventricular thrombus consistently present in isografts and present late (<10 days PTx) in allografts likely prevented appreciable loading volumes to cause ventricular remodelling, even in the face of high AR. As far as the changes in the cellular and biochemical profiles induced by LV unloading, research suggests that grafts undergo highly dynamic molecular and biochemical changes, reflecting the unique features that must be accounted for when studying unloaded HHTx models. These changes include: decreased cardiomyocyte size and volume; increased collagen concentration; increased matrix metalloproteinase (MMP-2 and -9) activity and zymographic activity; altered synthesis of cytoskeletal proteins; decreased enzymatic activity of myofibrillar adenosine diphosphatase, 6-phospho-fructokinase, citrate synthase, and other enzymes; and decreased total cardiac protein content (McGowan *et al.*, 2003; Spencer *et al.*, 2003).

Chamber dimensions have been used to identify acutely rejecting grafts both clinically (Dodd *et al.*, 1993) and in murine HHTx grafts (Scherrer-Crosbie *et al.*, 2002). Our experimental results suggest that despite a general trend for decreased dimensions early PTx, later increases (>10 days PTx) in EDD and ESD are prominent in allografts (**Fig. 14A, B**); this is in contrast to isografts, where LV dimensions remained generally stable throughout the observed period.

The most plausible explanation for the observed increases in LV diameters after 10 days PTx in allografts and not isografts may reflect changes brought about by the prevalence of aortic insufficiency (AI) and AR in this model system. In clinical transplantation, AI/AR can be caused by the presence of aortic valvular edema due to lymphocyte infiltration and inflammation (Oei *et al.*, 2000; Cladellas *et al.*, 1994) and widening of the aortic root, the latter being the most common aetiology - amounting to 80% of AI/AR cases (Cladellas *et al.*, 1994). AR also leads to increased SV and pre- and afterload volumes, eventually inducing LV remodelling manifested as increased LV diameters (Dujardin *et al.*, 1999a). Consequently, due to the increased loading volumes and pressures, human hearts decompensate and generation of edema and the presence of engorged blood vessels can be observed (Oei *et al.*, 2000). In our experiments, we observed analogous characteristics in

111

allografts, suggesting myocardial decompensation mediated by AR/AI is responsible for many of the changes in acutely rejecting allografts. For example, we observed that allografts show a gradual thickening of the aortic valves and aortic root widening as well as increased coronary vessel diameter from B-mode imaging (**Fig. 31C**), though these were not quantified. UBM imaging also verified the observation that LV cavity thrombi, which is ubiquitously present in all grafts, is dissolved and absent in allografts late after transplant (>10 days) (**Fig. 31C,D**), but not isografts (as thrombus is still present throughout), coinciding with the steep increase in LV diameters. The dissipation of LV cavity thrombus seen exclusively in allografts has not been previously described and may be the result of overall increased LV loading volumes (**Fig. 31A**) entering the graft cardiac cycle, which may be due to widening of the aortic root; the augmented volume could facilitate the breakdown of the fibrin network of coagulated blood. Taken together, persistently high AR (>1000 mm/s) (**Fig. 16B**) and the absence of LV cavity thrombus after day 10 PTx, may be the primary causes of LV remodelling late after transplant in allografts. There is a caveat in this explanation, however, as the manifestation of AI/AR - both clinically (Dujardin *et al.,* 1999b) and in experimental studies involving native hearts and graft rat HHTx grafts (McGowan *et al.,* 2003; Nakamura *et al.,* 2001) – appears to also stimulate LV hypertrophy, which is clearly the opposite trend seen late in atrophied allografts. In this regard, Spencer *at al.* have studied rat HHTx grafts where AI/AR was induced by mechanical aortic valvotomy and have concluded that moderate AR is insufficient to induce hypertrophy in this model system but may still lead to LV remodelling (Spencer *et al.,* 2003). The authors noted that LV remodelling and changes in LV compliance, but not LV hypertrophy, were sensitive to moderate AR (Spencer *et al.,* 2003); their echocardiographic patterns describing morphologic changes in regurgitant rat HHTx grafts also concur with our findings (i.e., increased EDD and ESD, decreased LV wall thickness relative to control transplants). Interestingly, a study by Bergsland *et al.* reported lower arterial pressure and CO in the native hearts of recipients receiving allografts in comparison to isografts (Bergsland *et al.,* 1989). This observation may be explained in light of the enlarging ventricular cavity of the allograft, which would allow more blood from the native circulation to enter the graft cardiac cycle and thus serve as a large conduit for recipient blood. The dilating allograft could potentially reduce venous pressure and consequently venous return and native heart CO. The persistent LV cavity thrombus present in isografts as well as low generation of edema and inflammation throughout the observed period may explain why they do not undergo similar changes in LV remodelling as allografts - despite high AR Doppler velocities (**Fig. 16B**) - since these factors effectively diminished AR-induced increases in load volume and consequently, prevented the isograft from undergoing ventricular remodelling.

4.6.3 Effects of Isoflurane Anaesthesia

It is important to note that measurements of ventricular dimensions have been reported to be affected by choice of anaesthetic agents (Yang *et al.*, 1999; Roth *et al.*, 2002). Yang *et al.* reported that, in addition to the negative chronotropic (decreased HR) and inotropic effects (reduced %FS and CO) of using pentobarbital and ketamine-xyline, measurements of ESD, EDD, septal and LV PWed were reduced when compared to echocardiographic data obtained from conscious mice. These cardiodepressive effects however were attenuated by using isoflurane, a popular choice of anaesthetic in small animal ultrasound imaging (and used in this study), which has been shown to yield the most reliable measurements of EDD and %FS in repeat studies on normal mouse hearts (Roth *et al.*, 2002). This is particularly relevant in this study since our experimental protocol utilized serial imaging of isoflurane-anesthetized mice. Though we are unaware of any data that has documented the effects of serial administration of isoflurane anaesthesia on cardiac function and murine echocardiographic parameters over the relatively long experimental period employed in this study, it appears that using isoflurane yields the most reproducible data in determining HR, %FS and EDD and brings about the least cardiodepressive effects when compared to pentobarbital or ketamine-xyline in normal mouse hearts during repeated echocardiographic examination 12 days apart (Roth *et al.*, 2002).

It is interesting to note that research into anaesthetic preconditioning has shown isoflurane to have possible cardioprotective effects and has been recently tested in dogs with occluded coronary arteries as well as in elderly patients undergoing coronary artery bypass graft surgery (Riess *et al.*, 2004; Kersten *et al.*, 1996). It appears that isoflurane can augment recovery of myocardial contractile function after reversible tissue injury produced by a 15-min left anterior descending coronary artery occlusion (LAD) and reperfusion in dogs (Kersten *et al.*, 1996). Whether inhaled isoflurane in our experimental protocol offered a

113

cardioprotective advantage following perioperative ischemia in murine cardiac grafts cannot be verified in this study, though in any case, it seems that graft viability and the median survival time of acutely rejecting allografts was not affected.

4.7   Doppler Flow

4.7.1   Graft Ventricular Pressure, LA Dilatation and MR

As early as day 8 PTx, MR Doppler velocities change dramatically (compare isografts: ~3000-4000 mm/s vs. allografts: >500mm/s, **Fig. 18B**). In non-transplanted human hearts, increased ventricular pressure as a result of AI/AR can lead to increased pressure in the left atrium (LA), which compensates by dilating (Braunwald, 1965; MEADOWS *et al.*, 1963). LA dilation has also been associated with secondary mitral regurgitation (MR) and mitral prolapse (Devereux *et al.*, 1976) – all observations verified either directly by UBM-Doppler or 2-D imaging of isografts. Though under-filling, ventricular atrophy, decreased ventricular dimensions, and the persistent presence of LV cavity thrombus in isografts throughout our experiments would be expected to attenuate volume and loading pressures substantially, Kolar *et al.* have noted that developed pressure and the maximum rate of pressure development were unaffected in atrophic rat HHTx ventricles (Kolar *et al.*, 1993). In addition, LV thrombus in isografts was observed in our experimental scheme to decrease in size over the observed period, and consequently, allowed for movement of a much more significant volume of ventricular blood in the vacated LV cavity space. In fact, LV thrombi decreased in size to the point that the clot was capable of traversing the aortic root during the cardiac cycle (**Fig. 31E, F**). Furthermore, findings from our previous study are in agreement with this premise when we observed consistently high AR, MR, LA dilatation, and %FS late after transplant in isografts, which we suggested involved the movement of a larger volume of blood flow through left heart and consequently, a larger change of LV volume throughout cardiac cycle in isografts (Zhou *et al.*, 2007). It appears that persistently high AR Doppler velocities, augmented cavity space and recipient circulatory blood volume entering the graft cardiac cycle - effects due to the shrinking thrombus, and the preserved ability for ventricular pressure development - are likely causes of the LA dilation and secondary MR seen in isografts. Though we did not directly measure ventricular pressure, LA dilation and MR (>8 days PTx) are close to nil in allografts and suggests that these two elements are not only

114

associated, but share a common cause in isografts, *i.e.*, higher ventricular pressure. Conversely, we suggest that ventricular pressure is diminished in acutely rejecting allografts, presumably due to ventricular remodelling, decreased compliance (loading) and systolic dysfunction as a result of ACR. Our explanation is also consistent with experiments performed on rat HHTx allografts, where peak systolic and diastolic LV pressures decreased significantly in rejecting day 5 PTx allografts, whereas they remained relatively unchanged in isografts (Szabo *et al.*, 2001).

Regarding the use of tricuspid inflow Doppler waveform Doppler data, our experiments confirmed results from our previous study, that the tricuspid and pulmonary waveforms were small in amplitude and variable in waveform, thus precluding these measurements as reliable parameters of evaluating rejection (Zhou *et al.*, 2007). Generally, interpretation of tricuspid diastolic inflow waveforms is questionable since the E and A-waves are not distinguishable in conscious mice and while anaesthesia may circumvent this problem, it casts doubt on the applicability and utility of these findings (Scherrer-Crosbie, 2006). The problem in rodent HHTx grafts is exacerbated by conditions of low pressure in the right heart, which was observed by UBM to result in right atrial and right ventricular collapse.

4.7.2 Ventricular Remodelling in Allografts and Increased Forward and Backward Aortic Flow Velocities

This is the first report that has characterized flow haemodynamics from the donor aorta - the major channel for supplying blood and perfusing the graft - in a rodent HHTx model. The observation that a generally higher backward (*i.e.*, downward flow away from the transducer) aortic flow velocity is seen in allografts (~200-400m/s) in comparison to isografts (~100-200 mm/s), a parameter which we have shown to be due to graft systole (**Figs. 19, 20**), would seem to contradict what one would expect from an acutely rejecting graft whose viability is diminishing. Decreased systolic function in acutely rejecting grafts was reported also in other studies in HHTx murine grafts, where a declining %EF and %FS is observed in allografts (Scherrer-Crosbie *et al.*, 2002). However, it is possible that, whereas inflammation, generation of pericardial fluid, a dissolving thrombus, and aortic root widening and consequent AR probably caused increases in LV cavity volume in allografts - which likely also induced ventricular remodelling - changes in maximal Doppler velocities

115

obtained from the ascending aorta seem to be associated with LV remodelling, illustrating an intriguing chain of events. Therefore, many of the same causes for increased ventricular dimensions may have acted also to increase LV blood volume entering the allograft cardiac cycle. These same phenomena consequently resulted in increased aortic forward and backward maximum velocities (**Figs. 21, 22**). Forward flow increased progressively post-transplant likely because LV pressures in allografts decreased (Soto *et al.*, 1998), allowing more blood to enter the donor aorta unimpeded from the recipient circulation. Conversely, maximal backward velocity and time-velocity integral reflect the increase in load volumes as a result of remodelling. This seems to indicate that aortic Doppler waveforms are heavily affected by morphological changes in the graft as a result of the generation of edema and inflammation.

4.7.3 Unique Doppler Waveforms and Decreased Coronary Flow Velocities Characterize Acutely Rejecting Allografts

As far as we are aware, this is the first report of experiments conducted to measure directly Doppler flow waveforms and velocities in the coronary arteries of a rodent HHTx model non-invasively using UBM. Measuring coronary flow was deemed important since graft perfusion is essential for graft viability (Kevin Wei, 2004) and severe impairment of coronary reserve has not only been associated with rejection in human patients with orthotopic heart transplants (Nitenberg *et al.*, 1989), but was an early predictor of LV dysfunction (Dandel *et al.*, 2003). Building on a previous study by our laboratories (Zhou *et al.*, 2007), forward coronary flow (**Fig. 23C**) was verified to be responsible for perfusion of graft myocardium and is likely due to native circulatory pressure whereas backward velocity waveforms are produced by graft cardiac systole (**Fig. 23D**). Similar to human native hearts, the majority of coronary (*i.e.*, positive/forward) flow in murine HHTx grafts occurs during graft diastole (Hozumi *et al.*, 1998); negative/backward waveforms, however, are produced during systole and this finding is corroborated by the absence of backward waveforms in severely rejected allografts, where contractility is significantly compromised. Thus, coronary waveforms of acutely rejected grafts (**Fig. 24B/28B**, *ALLO:D10*) are similar to the pulsatile and continuous flow seen in **Fig. 23C**, where flow is due only to recipient circulatory pressure. This suggests that poor graft viability of a typical day 10 allograft yields a unique Doppler spectrum that lacks waveforms due to graft ventricular contraction, namely the

116

absence of backward flow, with manifestation of a continuous, pulsatile flow pattern as a result of the native recipient cardiac cycle and peripheral blood pressure. That declining Bwd Vmax values for LCA (**Figs. 25, 26**) and RCA Doppler (**Figs. 29, 30**) spectra are observed in allografts after day 7 PTx suggests that severe impairment in myocardial contractility can be observed by assessing coronary waveform maximal velocities and integrals, and presents a novel, non-invasive method of evaluating graft viability in an unloaded model where traditional indices of graft function and contractility are not useful. Furthermore, reduced coronary perfusion - reflected by declining Fwd Vmax values - after day 5 for LCA and day 6 PTx for RCA waveforms, seems to reflect a consequent decline in overall contractility as illustrated by diminishing LCA and RCA Bwd Vmax values at day 7 PTx in both arteries, suggesting that adequate perfusion is vital for graft systolic ventricular function (Dandel *et al.*, 2003; Zhou *et al.*, 2007).

The ubiquitous observation of a recovery period during days 1-3 PTx from low to higher LCA and RCA Vmax and TVI values (**Figs. 24-26, 28-30**) is very likely due to I/R-induced trauma and data from both isografts and allografts are very similar during this period. The recovery period from I/R injury in rats HHTx grafts however, seems to be 2 days shorter than in mice (Szabo *et al.*, 2001), and our laboratory has previously shown that maximal forward velocity flow to perfuse the myocardium significantly increased from day 1 to 5 in isografts (Zhou *et al.*, 2007). Following the post-ischemic recovery phase, our data suggest that optimal myocardial perfusion seems to occur during 5-6 days PTx in murine allografts, whereas the pattern was more variable in isografts.

Previous experiments measuring coronary or myocardial blood flow in HHTx models have yielded variable results. Using the hydrogen clearance method, Szabo *et al.* observed a marked decrease in myocardial blood flow in 3 and 5 day PTx rat allografts in association with histological mild to moderate rejection, whereas it remained stable in isografts (Szabo *et al.*, 2001). While a previous report showed that coronary perfusion is not affected by ACR in canine HHTx models (Bando *et al.*, 1991), the majority of the literature would seem to contradict this premise (Hamano *et al.*, 1989; Libersan *et al.*, 1997; Bergsland *et al.*, 1989; Zhou *et al.*, 2007; Szabo *et al.*, 2001). Bergsland *et al.* noted low coronary flow in acutely rejecting day 6 rat HHTx allografts (Bergsland *et al.*, 1989) and Libersand *et al.*, several years after Bando *et. al*'s study, described restricted coronary perfusion during severe

117

rejection in dog HHTx grafts, which the authors suggested was due to ischemic and vascular heart disease (Libersan et al., 1997).

Our results show that maximum LCA and RCA forward velocities (**Fig. 25A, 29A**) in allografts peaked on days 4-5 and decreased gradually thereafter, though only RCA forward Vmax values decreased significantly in comparison to isografts. Because forward Vmax is generated by the native heart and is responsible for perfusing the graft, the decrease in forward flow velocity perhaps reflected changes as a result of ACR, atrophy, and low physiologic need for oxygen and nutrients in the severely diminished right ventricle (Rakusan et al., 1997; Cross et al., 1961). Furthermore, there is evidence to suggest that coronary artery disease, manifested as fibrointimal hyperplasia, is associated with increased mean ACR score (Alexander et al., 2005). It is therefore possible that luminal narrowing of coronary microvascular vessels augmented resistance to flow and a consequent decrease in forward and backward LCA and RCA flow velocities, though we did not measure flow in this study directly. In addition, it is likely that decreased coronary flow velocities are affected by the engorgement and widening of the coronary artery at the site of velocity measurement (**Fig. 31C,** unquantified observation), which would decrease both forward and backward flow velocities in allografts. Engorged blood vessels have been established clinically to be due to cardiac decompensation induced by interstitial edema and increased LV volumes in human hearts (Oei et al., 2000), and we suspected that similar mechanisms may be responsible for these observations in allografts (see above). Diminishing graft contraction strength likely contributes to the gradual decrease in backward flow, but would not be expected to affect forward flow. However, both forward and backward coronary flow velocities decreased in acutely rejecting allografts post-transplant. Thus, we speculate that a combination of factors, including increased vascular resistance due to coronary disease, widening of coronary vessel diameters, and diminishing systolic function in allografts, contributed to the decreased coronary flow velocities in the LCA and RCA. Further research is required to determine the primary mechanism(s) underlying these changes.

It should be noted that, due to the variable date of allografts 'falling out' from the data set, one should be cautioned in over-interpreting the lack of detectable differences in LCA forward blood flow velocity between isografts and allografts late after transplant. Because coronary perfusion is tightly coupled to oxygen consumption and hydrostatic pressures of microvascular resistance vessels within the myocardium (Klabunde, 2004), it is very possible that involution of capillaries in the atrophied ventricles of isografts and allografts would reflect decreased forward LCA and RCA flow (Rakusan *et al.*, 1997). This would seem to corroborate our previous study where a decrease in forward LCA flow from day 5 to 14 and stabilization of flow thereafter were reported in isografts (Zhou *et al.*, 2007). Nevertheless, in this model it would seem that in addition to microvascular resistance and oxygen requirements, one must account for the complex interaction of aortic and ventricular pressures from asynchronous native and graft cardiac cycles, as well as the ability for graft coronary vasculature to perform autoregulation of blood flow. These lines of inquiry present a fascinating venue for further research.

4.8 Limited Usefulness of Systolic Parameters of Graft Physiology

Our results indicated that due to the unloaded nature of the model used and the development of a large ventricular thrombus in the LV cavity (Scherrer-Crosbie *et al.*, 2002; Zhou *et al.*, 2007), the load-dependent variables that were measured (*i.e.*, %EF, %FS, CO, SV) were not reliable measures of graft physiology in heterotopic murine cardiac grafts. We have found traditional indices of systolic function to be highly variable in murine HHTx grafts, precluding generalizations about the trends in the PTx period. The limited information from systolic measures of LV function is echoed in the clinical arena as well (Mannaerts *et al.*, 1992), and has been shown not to be rejection-specific (Sade *et al.*, 2006). Nevertheless, we have found that late-transplant CO and SV values were significantly different from isografts and that generally, there seems to be an increasing trend for these measurements for allografts in the PTx period. It should be cautioned however that this apparent trend for increasing CO and SV values may by due to the effects of low graft ventricular pressure, interstitial edema (inducing aortic root widening), and the consequent, passive increase of native blood volume into the graft LV cavity. Therefore, these observations likely do not represent augmented systolic function.

Work by DiSesa *et al.* has substantiated the use of physiologic parameters in evaluating rodent cardiac grafts in reporting a significant decrease in CO and stroke work as being associated with acutely rejecting day 3 PTx rat HHTx grafts following isoproterenol treatment (DiSesa *et al.*, 1991). This finding may indicate that systolic parameters may be useful for only comparative measurements within a single treatment group and applicable when confined to specific time points and/or within a reasonably short time period. Indeed, the finding of diminishing trends in %EF and %FS from days 3 to 5 PTx in murine HHTx allografts (Scherrer-Crosbie *et al.*, 2002) substantiates this view. Kolar *et al.* also showed that systolic mechanical performance of rat HHTx cardiac isografts was maintained during the development of atrophy in isografts, enabling suitable comparisons with acutely rejecting grafts (Kolar *et al.*, 1993). Finally, our group has previously reported significant increases in %FS from days 14 versus 50 PTx in isografts as assessed by UBM-Doppler (Zhou *et al.*, 2007) and illustrates the fact that LV systolic indices of cardiac function may be used to evaluate short-term or long-term changes in cardiac physiology but are nevertheless, applicable only to specific time points and not likely to yield valuable information using serial evaluation.

4.9 Timetable of Echocardiographic and Histopathological Changes during Murine Heterotopic Acute Cardiac Graft Rejection

A general reference timetable based on our experimental findings and consideration of the results is summarized in **Table IX**. Possible causes and mechanisms based on the literature are given where applicable.

Table IX. Timetable of Echocardiographic and Histopathological Changes during Murine Heterotopic Acute Cardiac Graft Rejection

PHASE	POST-ISCHEMIC PHASE		MID-TRANSPLANT PHASE		LATE TRANSPLANT PHASE	
Corresponding Day Post-Transplant (PTx)	Days 1-3		Days 4-7		Days >8	
	ISOGRAFTS	ALLOGRAFTS	ISOGRAFTS	ALLOGRAFTS	ISOGRAFTS	ALLOGRAFTS
General Measurements						
Palpation	↔ ↓	↔ ↓	↔ ↓	↑ ↓	↔ ↓	↑ ↓
Primary etiology/mechanism(s)	I/R injury, Th formation	I/R injury, Th formation	—	ACR	—	ACR, LV atrophy
Graft Heart Rate (HR)	↔ ↓	↔ ↓	↔ ↓	↔ ↓	↔ ↓	↓ ↑
Primary etiology/mechanism(s)	I/R injury	I/R injury	—	ACR	—	ACR
Graft Weight	NR	NR	LV atrophy	Myocardial edema, ACR	LV atrophy	Myocardial edema, ACR
Primary etiology/mechanism(s)	—	—				
Cardiac Graft Morphology						
LV Wall Thicknesses	↔ ↓		↔ ↓	↔ ↑	LV atrophy	LV Atrophy
Primary etiology/mechanism(s)	—	I/R injury, Vascular permeability, Interstitial edema	—	ACR, Myocardial edema	—	—
LV Chamber Diameters	↔ ↓	↔ ↓	↔ ↓	↔ ↓	↔ ↓	Myocardial edema, AI/AR, ↑Blood volume, LV remodeling, ACR
Primary etiology/mechanism(s)	—	—	—	—	—	
LA Chamber Size	↔ ↑	↔ ↑	Th reduction, ↑Blood volume, AR/AI, ↓LV pressure, ↑LA pressure	↓LV pressure, ↓LA pressure	Th reduction, ↑Blood volume, AR/AI, ↑LV pressure, ↑LA pressure	↓LV pressure, ↓LA pressure, ACR
Primary etiology/mechanism(s)	—	—				
Haemodynamics						
Valvular Regurgitation	AR: ↔ ↑; MR: ↔ ↑	AR: ↔ ↑; MR: ↔ ↑	AR: ↔ ↑; MR: ↔ ↑	AR: ↔ ↑; MR: ↔ ↑	AR: ↓; MR: ↑; ↑LA pressure, LA dilation, ACR	AR: ↓; MR: ↓; ↓LV pressure, ↓LA pressure, ACR
Primary etiology/mechanism(s)	—	—	—	—		
Aortic Flow	↔ ↑	↔ ↑	↔ ↑	Myocardial edema, LV remodeling	↔ ↑	Myocardial edema, AI/AR, ↑Blood volume, LV remodeling, ACR
Primary etiology/mechanism(s)	—	—	—		—	
Coronary Perfusion	Recovering from I/R injury	Recovering from I/R injury	↔ ↑	↔ ↑	↔ ↑	ACR, Ischemic vascular disease
Primary etiology/mechanism(s)	↔ ↑; —		—	—	—	
Histopathology						
Acute Cellular Rejection		Mild		Moderate		Severe
Primary etiology/mechanism(s)		Mild lymphocyte infiltration		Multi-focal lymphocyte infiltration, Cardiomyocyte necrosis		Widespread lymphocyte infiltration, Hemorrhage and/or vasculitis, Cardiomyocyte necrosis
Apoptosis (and associated echogenicity)	Apoptosis: ↑; Echogenicity: ↑	Apoptosis: ↑; Echogenicity: ↑	Apoptosis: NR; Echogenicity: ↑↑	Apoptosis: ↑; Echogenicity: ↑↑	Apoptosis: ↑↑; Echogenicity: ↑↑	Apoptosis: ↑↑; Echogenicity: ↑↑
Primary etiology/mechanism(s)	—	CTL-induced apoptosis, Subcellular nuclear changes, Cellular fragmentation	—	CTL-induced apoptosis, Subcellular nuclear changes, Cellular fragmentation	—	CTL-induced apoptosis, Subcellular nuclear changes, Cellular fragmentation

Abbreviations: NR, Not reported; I/R, ischemia-reperfusion; ACR, acute cardiac rejection; BPM, beats per minute; LV, left ventricle; Th, left ventricular thrombus; MR, mitral regurgitation; AR, aortic regurgitation; CTL, cytotoxic T cells

4.10   Study Limitations

In our attempt to thoroughly characterize HHTX murine grafts non-invasively using UBM, we did not directly evaluate diastolic indices of graft function. However, most of the methods that have been used to assess diastolic dysfunction in rejecting rodent grafts (*e.g.,* relaxation time constant, dP/d*t*max, LV diastolic pressure) have been procured invasively using implanted devices in the graft as an adjunct to echocardiographic parameters. It is also likely that invasive procedures will need to be utilized to evaluate adequately ventricular systolic function in this 'non-working model' system. We also did not directly associate UBM-derived data with the degree of histological damage due to ACR and apoptosis during the rejection episode. Nevertheless, we have undertaken to provide a fuller, more integrated approach to understanding a typical time-course in this model by compartmentalizing major changes as imaged by UBM imaging with widely reported histopathological changes. In addition, a more objective method for determining increases in backscatter would give greater clarity as to the specific changes that occur post-transplant and whether these changes are due to cellular processes as a result of apoptosis or other mechanisms. Finally, the calculated intra- and interobserver variabilities in this study were limited by the small number of observers used and therefore, results cannot be generalized to provide an indication of the variability of the method *per se.*

4.11 Conclusions

In conclusion, using the UBM, we detected differences between acutely rejecting allografts and accepted isografts earlier than the widely used technique of finger palpation in this experimental model. In particular, whereas rejecting grafts were first detected on day 7 by palpation, UBM-derived LV anterior and posterior wall thicknesses were significantly increased as early as day 2 PTx. In addition, graft HR was significantly reduced from day 7 onwards, aortic backward and forward velocities were significantly increased from days 6 and 8, respectively, and late changes in the flow velocities of the left and right coronary arteries, mitral regurgitation, and chamber dimensions were also observed. Decreased forward and backward coronary flow velocities in allografts may reflect a combination of factors, such as increased downstream resistance due to stenosing, vasculopathic coronary

122

vessels, engorgement of coronary vessels due to myocardial decompensation, and diminished allograft contractility in response in immunologic rejection.

Furthermore, our method yielded additional valuable information not attainable by finger palpation or other common methods, such as histology. In particular, we generated a comprehensive description of physiologic, morphologic, and haemodynamic changes using a novel technology on a common model of experimental cardiac transplantation during a course of graft rejection. The high spatial resolution of the cutting-edge UBM system has allowed investigators, for the first time, to describe intracardiac structures, ventricular dimensions, and unique Doppler flow waveforms in the small murine heart and we have adapted this technology to establish a technique for quick, non-invasive, serial characterization of murine HHTx grafts to document changes in the post-transplant period. The dynamic morphological and haemodynamic observations procured by M-mode, B-mode, and Doppler imaging reflect the changes that take place within a short period after transplantation of rodent grafts, verifying the need to assess critically the natural course of graft rejection in rodent models before genetic and pharmacological studies are performed in this model system.

4.12   Future Directions

There are several lines of further inquiry that may result from this work.

1. To describe how UBM-derived parameters change in murine studies in which graft donors and/or recipients are treated with immunosuppressive agents or tolerizing inocula (*e.g.,* fetal liver cells) or in which transgenic and/or knock-out mice have been used to investigate immunological modulation and graft acceptance. A comparison of how certain molecular, cellular, and pharmacological factors affect the physiology, morphology and haemodynamic profiles of murine HHTx grafts as assessed by UBM may more clearly delineate the mechanisms by which these variables act to induce tolerance, delay rejection, or inhibit inflammatory processes characteristic of graft rejection. Furthermore, investigators could explain changes in the physiology, haemodynamic and morphological of heterotopic grafts following such treatments more clearly in light of an established timeline of documented events in this model system, such as the one we have described.

123

2. Determining modified echo-derived parameters of physiologic function seeing that traditional indices of systolic performance yield limited information and current diastolic parameters are procured invasively.

3. Documenting echocardiographic changes from murine HHTx models of different genetic disparities and patterns of rejection, such as hyperacute, acute vascular, or chronic rejection/graft vasculopathy, using established strain combinations that give rise to these rejection patterns.

V. REFERENCES

"Canada's Heart Transplantation Rate Down; Survival Rates Up." CIHI Reports. 2004

Canadian Organ Replacement Registry Data. www.chi.ca. 2006.

Abbott,C.P., Creech,O., Jr., and Dewitt,C.W. (1964a). Histologic and electrocardiographic changes of the transplanted rat heart. Surg Forum. *15:253-5.*, 253-255.

Abbott C.P., Dewitt,C.W., and Creech,O., Jr. (1965). The transplanted rat heart: histologic and electrocardiographic changes. Transplantation. *3:432-45.*, 432-445.

Abbott,C.P., Lindsey,E.S., Creech,O., Jr., and Dewitt,C.W. (1964b). A technique for heart transplantation in the rat. Arch. Surg. *89:645-52.*, 645-652.

Affleck,D.G., Bull,D.A., Albanil,A., Shao,Y., Brady,J., Karwande,S.V., Eichwald,E.J., and Shelby,J. (2001). Interleukin-18 production following murine cardiac transplantation: correlation with histologic rejection and the induction of INF-gamma. J Interferon Cytokine Res. *21*, 1-9.

Akashi,S., Sho,M., Kashizuka,H., Hamada,K., Ikeda,N., Kuzumoto,Y., Tsurui,Y., Nomi,T., Mizuno,T., Kanehiro,H., Hisanaga,M., Ko,S., and Nakajima,Y. (2005). A novel small-molecule compound targeting CCR5 and CXCR3 prevents acute and chronic allograft rejection. Transplantation *80*, 378-384.

Akosah,K.O., Olsovsky,M., Kirchberg,D., Salter,D., and Mohanty,P.K. (1996). Dobutamine stress echocardiography predicts cardiac events in heart transplant patients. Circulation. *94*, II283-II288.

Alexander,D.Z., Pearson,T.C., Hendrix,R., Ritchie,S.C., and Larsen,C.P. (1996). Analysis of effector mechanisms in murine cardiac allograft rejection. Transpl. Immunol. *4*, 46-48.

Alexander,R.T., Lathrop,S., Vollmer,R., Blue,L., Russell,S.D., and Steenbergen,C. (2005). Graft vascular disease after cardiac transplantation and its relationship to mean acute rejection score. Arch. Pathol Lab Med. *129*, 1283-1287.

Alpert,S., Lewis,N.P., Ross,H., Fowler,M., and Valantine,H.A. (1995). The relationship of granzyme A and perforin expression to cardiac allograft rejection and dysfunction. Transplantation *60*, 1478-1485.

Alyono,D., Crumbley,A.J., III, Schneider,J.R., Bolman,R.M., III, Chao,R.Y., McGregor,L., and Anderson,R.W. (1984). Early mechanical function in the heterotopic heart transplant. J Surg. Res *37*, 55-62.

Amende,I., Simon,R., Seegers,A., Daniel,W., Heublein,B., Hetzer,R., Haverich,A., Hood,W.P., Jr., Lichtlen,P.R., Schutzenmeister,R., and . (1990). Diastolic dysfunction during acute cardiac allograft rejection. Circulation. *81*, III66-III70.

Amirhamzeh,M.M., Jia,C.X., Starr,J.P., Sciacca,R., Chowdhury,N.C., Hsu,D.T., and Spotnitz,H.M. (1994). Diastolic function in the heterotopic rat heart transplant model. Effects of edema, ischemia, and rejection. J Thorac Cardiovasc Surg. *108*, 928-937.

Anderson,C.A., Shernan,S.K., Leacche,M., Rawn,J.D., Paul,S., Mihaljevic,T., Jarcho,J.A., Stevenson,L.W., Fang,J.C., Lewis,E.F., Couper,G.S., Mudge,G.H., and Byrne,J.G. (2004). Severity of intraoperative tricuspid regurgitation predicts poor late survival following cardiac transplantation. Ann. Thorac. Surg. *78*, 1635-1642.

Angermann,C.E., Nassau,K., Stempfle,H.U., Kruger,T.M., Drewello,R., Junge,R., Uberfuhr,P., Weiss,M., and Theisen,K. (1997). Recognition of acute cardiac allograft rejection from serial integrated backscatter analyses in human orthotopic heart transplant recipients. Comparison with conventional echocardiography. Circulation *95*, 140-150.

Arras,M., Autenried,P., Rettich,A., Spaeni,D., and Rulicke,T. (2001). Optimization of intraperitoneal injection anesthesia in mice: drugs, dosages, adverse effects, and anesthesia depth. Comp Med. *51*, 443-456.

Asante-Korang,A. (2004). Echocardiographic evaluation before and after cardiac transplantation. Cardiol Young. *14 Suppl 1:88-92.*, 88-92.

Bando,K., Fraser,C.D., Jr., Chacko,V.P., Pillai,R., Jacobus,W.E., Cameron,D.E., Hutchins,G.M., Reitz,B.A., and Baumgartner,W.A. (1991). Coronary blood flow does not decrease during allograft rejection in heterotopic heart transplants. J Heart Lung Transplant. *10*, 251-257.

Barbee,R.W., Perry,B.D., Re,R.N., and Murgo,J.P. (1992). Microsphere and dilution techniques for the determination of blood flows and volumes in conscious mice. Am J Physiol. *263*, R728-R733.

Baxter, D and Smerdon, J. Donation Matters: Demographics and Organ Transplants in Canada, 2000 to 2040. 46. 2000. Urban Future Institute.

Bergese,S.D., Klenotic,S.M., Wakely,M.E., Sedmak,D.D., and Orosz,C.G. (1997). Apoptosis in murine cardiac grafts. Transplantation. *63*, 320-325.

Bergsland,J., Hwang,K., Driscoll,R., Carr,E.A., Wright,J.R., Curran-Everett,D.C., Carroll,M., Krasney,E., and Krasney,J.A. (1989). Coronary blood flow and thallium 201 uptake in rejecting rat heart transplantations. J Heart Transplant. *8*, 147-153.

Bernhard,A. and Konertz,W. (1983). Experimental heart transplantation. J Thorac Cardiovasc Surg. *86*, 314-315.

Berson,M., Vaillant,L., Patat,F., and Pourcelot,L. (1992). High-resolution real-time ultrasonic scanner. Ultrasound Med. Biol. *18*, 471-478.

Bhatia,S.J., Kirshenbaum,J.M., Shemin,R.J., Cohn,L.H., Collins,J.J., Di Sesa,V.J., Young,P.J., Mudge,G.H., Jr., and Sutton,M.G. (1987). Time course of resolution of pulmonary hypertension and right ventricular remodeling after orthotopic cardiac transplantation. Circulation. *76*, 819-826.

Billingham,M.E. (1992). Histopathology of graft coronary disease. J Heart Lung Transplant. *11*, S38-S44.

Billingham,M.E., Caves,P.K., Dong E Jr, and Shumway,N.E. (1973). The diagnosis of canine orthotopic cardiac allograft rejection by transvenous endomyocardial biopsy. Transplant Proc. *5*, 741-743.

Bishop S. (1980). Cardiovascular Research. (New York: Academic Press), p. 161.

Blake,M.H. (2004). Mechanisms of Acute Transplant Rejection. South Carolina Journal of Molecular Medicine 109-116.

Bom,N., ten Hoff,H., Lancee,C.T., Gussenhoven,W.J., and Bosch,J.G. (1989). Early and recent intraluminal ultrasound devices. Int. J Card Imaging. *4*, 79-88.

Bonham,C.A., Peng,L., Liang,X., Chen,Z., Wang,L., Ma,L., Hackstein,H., Robbins,P.D., Thomson,A.W., Fung,J.J., Qian,S., and Lu,L. (2002). Marked prolongation of cardiac allograft survival by dendritic cells genetically engineered with NF-kappa B oligodeoxyribonucleotide decoys and adenoviral vectors encoding CTLA4-Ig. J Immunol. *169*, 3382-3391.

Boucek,M.M., Mathis,C.M., Kanakriyeh,M.S., Hodgkin,D.D., Boucek,R.J., Jr., and Bailey,L.L. (1993). Serial echocardiographic evaluation of cardiac graft rejection after infant heart transplantation. J Heart Lung Transplant. *12*, 824-831.

Bourge,R.C., Naftel,D.C., Costanzo-Nordin,M.R., Kirklin,J.K., Young,J.B., Kubo,S.H., Olivari,M.T., and Kasper,E.K. (1993). Pretransplantation risk factors for death after heart transplantation: a multiinstitutional study. The Transplant Cardiologists Research Database Group. J Heart Lung Transplant. *12*, 549-562.

Boyd,S.Y., Mego,D.M., Khan,N.A., Rubal,B.J., and Gilbert,T.M. (1997). Doppler echocardiography in cardiac transplant patients: allograft rejection and its relationship to diastolic function. J Am Soc. Echocardiogr. *10*, 526-531.

Braunwald,E. (1965). Pathologic physiology of valvular regurgitation. Verh. Dtsch. Ges. Kreislaufforsch. *31:36-50.*, 36-50.

Burdick,J.F. and Clow,L.W. (1986). Rejection of murine cardiac allografts. I. Relative roles of major and minor antigens. Transplantation *42*, 67-72.

Burgess,M.I., Bhattacharyya,A., and Ray,S.G. (2002). Echocardiography after cardiac transplantation. J Am Soc. Echocardiogr. *15*, 917-925.

Bushong,S.C. and Benjamin,R.A. (1991). **Diagnostic ultrasound: Physics, biology, and instrumentation**. (St. Louis: Mosby Year Book).

Campos,L., Deli,B.C., Kern,J.H., Kim,J.I., Naji,A., Barker,C.F., and Markmann,J.F. (1995). Survival of MHC deficient mouse heterotopic cardiac allografts and xenografts. Transplant Proc. *27*, 254-255.

Caves,P.K., Stinson,E.B., Billingham,M.E., Rider,A.K., and Shumway,N.E. (1973). Diagnosis of human cardiac allograft rejection by serial cardiac biopsy. J Thorac Cardiovasc Surg. *66*, 461-466.

Chandrasekaran,K., Bansal,R.C., Greenleaf,J.F., Hauck,A., Seward,J.B., Tajik,A.J., and Bailey,L.L. (1987). Early recognition of heart transplant rejection by backscatter analysis from serial 2D ECHOs in a heterotopic transplant model. J Heart Transplant. *6*, 1-7.

Cho,S.I., Marcus,F.S., and Kountz,S.L. (1972). A new model for study of allograft rejection in the rat: use of skin with an intact vascular pedicle. I. Transplantation. *13*, 486-492.

Christopher,K., Mueller,T.F., DeFina,R., Liang,Y., Zhang,J., Gentleman,R., and Perkins,D.L. (2003). The graft response to transplantation: a gene expression profile analysis. Physiol Genomics. *15*, 52-64.

Cladellas,M., Oriol,A., and Caralps,J.M. (1994). Quantitative assessment of valvular function after cardiac transplantation by pulsed Doppler echocardiography. Am J Cardiol. *73*, 1197-1201.

Collins,K.A., Korcarz,C.E., and Lang,R.M. (2003). Use of echocardiography for the phenotypic assessment of genetically altered mice. Physiol. Genomics *13*, 227-239.

Cooper,D.K., Keogh,A.M., Brink,J., Corris,P.A., Klepetko,W., Pierson,R.N., Schmoeckel,M., Shirakura,R., and Warner,S.L. (2000). Report of the Xenotransplantation Advisory Committee of the International Society for Heart and Lung Transplantation: the present status of xenotransplantation and its potential role in the treatment of end-stage cardiac and pulmonary diseases. J Heart Lung Transplant. *19*, 1125-1165.

Corry,R.J., Winn,H.J., and Russell,P.S. (1973a). Heart transplantation in congenic strains of mice. Transplant Proc. *5*, 733-735.

Corry,R.J., Winn,H.J., and Russell,P.S. (1973b). Primarily vascularized allografts of hearts in mice. The role of H-2D, H-2K, and non-H-2 antigens in rejection. Transplantation *16*, 343-350.

Cramer,D.V., Chapman,F.A., Jaffee,B.D., Eiras-Hreha,G., Yasunaga,C., Wu,G.D., and Makowka,L. (1992a). The effect of a new immunosuppressive drug, brequinar sodium, on concordant hamster-to-rat cardiac xenografts. Transplant Proc. *24*, 720-721.

Cramer,D.V., Chapman,F.A., Jaffee,B.D., Jones,E.A., Knoop,M., Hreha-Eiras,G., and Makowka,L. (1992b). The effect of a new immunosuppressive drug, brequinar sodium, on heart, liver, and kidney allograft rejection in the rat. Transplantation. *53*, 303-308.

Cramer,D.V., Podesta,L.G., and Makowska,L. (1994). **Handbook of Animal Models in Transplantation Research.** (Boca Raton, Florida, USA: CRC Press Inc.).

Cross,C.E., Rieben,P.A., and Salisbury,P.F. (1961). Coronary driving pressure and vasomotor tonus as determinants of coronary blood flow. Circ Res *9*, 589-600.

Curie P and Curie J (1880). Developpement, par pression de l'electricite polaire dans les cristaux hemiedres a faces inclinees. Comptes Rendus. . *91*, 291-295.

Czarnota,G.J., Kolios,M.C., Abraham,J., Portnoy,M., Ottensmeyer,F.P., Hunt,J.W., and Sherar,M.D. (1999). Ultrasound imaging of apoptosis: high-resolution non-invasive monitoring of programmed cell death in vitro, in situ and in vivo. Br. J Cancer. *81*, 520-527.

Czarnota,G.J., Kolios,M.C., Vaziri,H., Benchimol,S., Ottensmeyer,F.P., Sherar,M.D., and Hunt,J.W. (1997). Ultrasonic biomicroscopy of viable, dead and apoptotic cells. Ultrasound Med. Biol. *23*, 961-965.

Dambrin,C., El Feghaly,M., Abbal,M., Glock,Y., Durand,D., Fournial,G., Ohayon,E., and Cerene,A. (1999). A new rejection criteria in the heterotopically placed rat heart by non-invasive measurement of Dp/Dtmax. J Heart Lung Transplant. *18*, 524-531.

Dandel,M., Wellnhofer,E., Hummel,M., Meyer,R., Lehmkuhl,H., and Hetzer,R. (2003). Early detection of left ventricular dysfunction related to transplant coronary artery disease. J Heart Lung Transplant *22*, 1353-1364.

Dawson,D., Lygate,C.A., Saunders,J., Schneider,J.E., Ye,X., Hulbert,K., Noble,J.A., and Neubauer,S. (2004). Quantitative 3-Dimensional Echocardiography for Accurate and Rapid Cardiac Phenotype Characterization in Mice. Circulation *110*, 1632-1637.

Del Castillo,J.M., Marotta,R.H., Ortiz,J., Matsumoto,A.Y., Souza,E.L., Silva,C.E., Macruz,R., and Carvalho,V.B. (1989). [Doppler echocardiographic evaluation of heart transplantation]. Arq Bras. Cardiol. *53*, 151-155.

Dengler,T.J. and Pober,J.S. (2000). Cellular and molecular biology of cardiac transplant rejection. J Nucl. Cardiol. *7*, 669-685.

Desruennes,M., Corcos,T., Cabrol,A., Gandjbakhch,I., Pavie,A., Leger,P., Eugene,M., Bors,V., and Cabrol,C. (1988). Doppler echocardiography for the diagnosis of acute cardiac allograft rejection. J Am Coll Cardiol. *12*, 63-70.

Devereux,R.B., Perloff,J.K., Reichek,N., and Josephson,M.E. (1976). Mitral valve prolapse. Circulation. *54*, 3-14.

DiSesa,V.J., Masetti,P., Diaco,M., Schoen,F.J., Marsh,J.D., and Cohn,L.H. (1991). The mechanism of heart failure caused by cardiac allograft rejection. J Thorac. Cardiovasc. Surg. *101*, 446-449.

Dodd,D.A., Brady,L.D., Carden,K.A., Frist,W.H., Boucek,M.M., and Boucek,R.J., Jr. (1993). Pattern of echocardiographic abnormalities with acute cardiac allograft rejection in adults: correlation with endomyocardial biopsy. J Heart Lung Transplant. *12*, 1009-1017.

Dodd,S.J., Williams,M., Suhan,J.P., Williams,D.S., Koretsky,A.P., and Ho,C. (1999). Detection of single mammalian cells by high-resolution magnetic resonance imaging. Biophys. J. *76*, 103-109.

Drolet,M.C., Roussel,E., Deshaies,Y., Couet,J., and Arsenault,M. (2006). A high fat/high carbohydrate diet induces aortic valve disease in C57BL/6J mice. J Am Coll Cardiol. *47*, 850-855.

Ducharme,A., Demers,P., Elkouri,S., Courval,J.F., Cartier,R., and Tardif,J.C. (1999). Characterization of the natural history of cervical heterotopic heart transplantation with echocardiography. J Heart Lung Transplant *18*, 510-516.

Dujardin,K.S., Enriquez-Sarano,M., Schaff,H.V., Bailey,K.R., Seward,J.B., and Tajik,A.J. (1999a). Mortality and morbidity of aortic regurgitation in clinical practice. A long-term follow-up study. Circulation. *99*, 1851-1857.

Dujardin,K.S., Enriquez-Sarano,M., Schaff,H.V., Bailey,K.R., Seward,J.B., and Tajik,A.J. (1999b). Mortality and Morbidity of Aortic Regurgitation in Clinical Practice : A Long-Term Follow-Up Study. Circulation *99*, 1851-1857.

Edler I (1956). Ultrasound cardiogram in mitral valve disease. Acta Chir Scand. *111*, 230.

Edler,I. and Hertz,C.H. (1954). The use of ultrasonic reflectoscope for the continuous recording of the movements of heart walls. Clin. Physiol Funct. Imaging. *24*, 118-136.

Effert,S. and Domanig,E. (1959). The diagnosis of intraatrial tumor and thrombi by the ultrasonic echo method. Med Meth. *4*.

El Sawy,T., Miura,M., and Fairchild,R. (2004). Early T Cell Response to Allografts Occuring Prior to Alloantigen Priming Up-Regulates Innate-Mediated Inflammation and Graft Necrosis. American Journal of Pathology *165*, 147-157.

Emanueli,C. and Madeddu,P. (2001). Angiogenesis gene therapy to rescue ischaemic tissues: achievements and future directions. Br. J Pharmacol. *133*, 951-958.

Fan,X., Ang,A., Pollock-Barziv,S.M., Dipchand,A.I., Ruiz,P., Wilson,G., Platt,J.L., and West,L.J. (2004). Donor-specific B-cell tolerance after ABO-incompatible infant heart transplantation. Nat. Med. *10*, 1227-1233.

Fan,X., Tyerman,K., Ang,A., Koo,K., Parameswaran,K., Tao,K., Mai,L., Lang,H., and West,L.J. (2005). A novel tool for B-cell tolerance research: characterization of mouse alloantibody development using a simple and reliable cellular ELISA technique. Transplant Proc. *37*, 29-31.

Fentzke,R.C., Korcarz,C.E., Shroff,S.G., Lin,H., Leiden,J.M., and Lang,R.M. (2001). The left ventricular stress-velocity relation in transgenic mice expressing a dominant negative CREB transgene in the heart. J Am Soc. Echocardiogr. *14*, 209-218.

Filipponi,F., Michel,A., and Houssin,D. (1989). Prolongation of guinea pig-to-rat xenograft survival with BN 52063, a specific antagonist of platelet-activating factor. Ital. J Surg Sci. *19*, 325-329.

Foster,F.S., Lockwood,G.R., Ryan,L.K., Harasiewicz,K.A., Berube,L., and Rauth,A.M. (1993). Principles and applications of ultrasound backscatter microscopy. Ultrasonics, Ferroelectrics and Frequency Control, IEEE Transactions on *40*, 608-617.

Foster,F.S., Pavlin,C.J., Harasiewicz,K.A., Christopher,D.A., and Turnbull,D.H. (2000). Advances in ultrasound biomicroscopy. Ultrasound Med. Biol. *26*, 1-27.

Furniss,S.S., Murray,A., Hunter,S., Dougenis,V., and McGregor,C.G. (1991). Value of echocardiographic determination of isovolumic relaxation time in the detection of heart transplant rejection. J Heart Lung Transplant *10*, 557-561.

Galinanes,M. and Hearse,D.J. (1991). Metabolic, functional, and histologic characterization of the heterotopically transplanted rat heart when used as a model for the study of long-term recovery from global ischemia. J Heart Lung Transplant. *10*, 79-91.

Geisterfer-Lowrance,A.A., Christe,M., Conner,D.A., Ingwall,J.S., Schoen,F.J., Seidman,C.E., and Seidman,J.G. (1996). A mouse model of familial hypertrophic cardiomyopathy. Science. *272*, 731-734.

Goertz,D.E., Needles,A., Burns,P.N., and Foster,F.S. (2005). High-frequency, nonlinear flow imaging of microbubble contrast agents. IEEE Trans. Ultrason. Ferroelectr. Freq. Control. *52*, 495-502.

Goldberg,B.B. (1991). Nonvascular endoluminal ultrasound makes debut. Diagn. Imaging (San. Franc.). *13*, 100-104.

Gottlieb,R.A., Burleson,K.O., Kloner,R.A., Babior,B.M., and Engler,R.L. (1994). Reperfusion injury induces apoptosis in rabbit cardiomyocytes. J Clin. Invest. *94*, 1621-1628.

Gould,D.S. and Auchincloss,H., Jr. (1999). Direct and indirect recognition: the role of MHC antigens in graft rejection. Immunol Today. *20*, 77-82.

Grasl-Kraupp,B., Ruttkay-Nedecky,B., Koudelka,H., Bukowska,K., Bursch,W., and Schulte-Hermann,R. (1995). In situ detection of fragmented DNA (TUNEL assay) fails to discriminate among apoptosis, necrosis, and autolytic cell death: a cautionary note. Hepatology. *21*, 1465-1468.

132

Grazia,T.J., Pietra,B.A., Johnson,Z.A., Kelly,B.P., Plenter,R.J., and Gill,R.G. (2004). A two-step model of acute CD4 T-cell mediated cardiac allograft rejection. J Immunol. *172*, 7451-7458.

Greenberg,M.L., Uretsky,B.F., Reddy,P.S., Bernstein,R.L., Griffith,B.P., Hardesty,R.L., Thompson,M.E., and Bahnson,H.T. (1985). Long-term hemodynamic follow-up of cardiac transplant patients treated with cyclosporine and prednisone. Circulation. *71*, 487-494.

Griepp,R.B., Stinson,E.B., Dong E Jr, Clark,D.A., and Shumway,N.E. (1971). Acute rejection of the allogrfted human heart. Ann. Thorac Surg. *12*, 113-126.

Griffiths,G.M. and Mueller,C. (1991). Expression of perforin and granzymes in vivo: potential diagnostic markers for activated cytotoxic cells. Immunol Today. *12*, 415-419.

Gronros,J., Wikstrom,J., Hagg,U., Wandt,B., and Gan,L.M. (2006). Proximal to middle left coronary artery flow velocity ratio, as assessed using color Doppler echocardiography, predicts coronary artery atherosclerosis in mice. Arterioscler. Thromb. Vasc. Biol. *26*, 1126-1131.

Hamano,K., Ohmi,M., Esato,K., and Mohri,H. (1989). Myocardial tissue blood flow in allotransplanted rat heart with a special reference to acute rejection. J Heart Transplant. *8*, 48-52.

Harland,C.C., Bamber,J.C., Gusterson,B.A., and Mortimer,P.S. (1993). High frequency, high resolution B-scan ultrasound in the assessment of skin tumours. Br. J Dermatol. *128*, 525-532.

Hart,C.Y., Burnett,J.C., Jr., and Redfield,M.M. (2001). Effects of avertin versus xylazine-ketamine anesthesia on cardiac function in normal mice. Am J Physiol Heart Circ Physiol. *281*, H1938-H1945.

Hau,J. and Van Hoosier,G.L. (2003). Handbook of Laboratory Animal Science. (Boca Raton, Florida, USA: CRC Press Inc.).

Haug,C.E., Shapiro,J.I., Chan,L., and Weil,R., III (1987). P-31 nuclear magnetic resonance spectroscopic evaluation of heterotopic cardiac allograft rejection in the rat. Transplantation. *44*, 175-178.

Healy,D.G., Watson,R.W., O'Keane,C., Egan,J.J., McCarthy,J.F., Hurley,J., Fitzpatrick,J., and Wood,A.E. (2006). Neutrophil transendothelial migration potential predicts rejection severity in human cardiac transplantation. Eur. J Cardiothorac. Surg. *29*, 760-766.

Heron,I. (1971). A technique for accessory cervical heart transplantation in rabbits and rats. Acta Pathol Microbiol. Scand. [A]. *79*, 366-372.

Heron,I. (1972). The iso--and allotransplanted rat heart. Histological, electrocardiographic and serological observations. Acta Pathol Microbiol. Scand. [A]. *80*, 9-16.

Hete, B., Shung, K. K., and Campbell, D. B. Cardiac allograft rejection monitored with ultrasound integrated backscatter. 41-42. 1990.

Hicks,M., Hing,A., Gao,L., Ryan,J., and Macdonald,P.S. (2006). Organ preservation. Methods Mol. Biol. *333:331-74.*, 331-374.

Higuchi,M. and Aggarwal,B.B. (1994). Differential roles of two types of the TNF receptor in TNF-induced cytotoxicity, DNA fragmentation, and differentiation. J Immunol. *152*, 4017-4025.

Hoffmann,K., el Gammal,S., and Altmeyer,P. (1990). [B-scan ultrasound in dermatology]. Hautarzt. *41*, W7-16.

Hoffmann,K., el Gammal,S., Matthes,U., and Altmeyer,P. (1989). [Digital 20 mhz sonography of the skin in preoperative diagnosis]. Z Hautkr. *64*, 851-858.

Hofmann,B., Tao,K., Mai,L., and West,L.J. (2004). Acceptance of related and unrelated cardiac allografts in neonatally tolerized mice is cardio-specific and transferable by regulatory CD4+ T cells. J Heart Lung Transplant. *23*, 1069-1076.

Hoit,B.D., Khoury,S.F., Kranias,E.G., Ball,N., and Walsh,R.A. (1995). In vivo echocardiographic detection of enhanced left ventricular function in gene-targeted mice with phospholamban deficiency. Circ Res. *77*, 632-637.

Hollenberg,S.M., Klein,L.W., Parrillo,J.E., Scherer,M., Burns,D., Tamburro,P., Bromet,D., Satran,A., and Costanzo,M.R. (2004). Changes in coronary endothelial function predict progression of allograft vasculopathy after heart transplantation. J Heart Lung Transplant *23*, 265-271.

Honjo,K., Xu,X., and Bucy,R.P. (2004). CD4+ T-cell receptor transgenic T cells alone can reject vascularized heart transplants through the indirect pathway of alloantigen recognition. Transplantation. *77*, 452-455.

134

Hosenpud,J.D., Bennett,L.E., Keck,B.M., Boucek,M.M., and Novick,R.J. (2001). The Registry of the International Society for Heart and Lung Transplantation: eighteenth Official Report-2001. J Heart Lung Transplant. *20*, 805-815.

Hozumi,T., Yoshida,K., Akasaka,T., Asami,Y., Ogata,Y., Takagi,T., Kaji,S., Kawamoto,T., Ueda,Y., and Morioka,S. (1998). Noninvasive assessment of coronary flow velocity and coronary flow velocity reserve in the left anterior descending coronary artery by Doppler echocardiography: Comparison with invasive technique. Journal of the American College of Cardiology *32*, 1251-1259.

Hunt,S. (2001). Reinnervation of the Transplanted Heart -- Why Is It Important? N Engl J Med *345*, 762-764.

Igal A.Sebag. Quantitative Assessment of Regional Myocardial Function in Mice by Tissue Doppler Imaging: Comparison With Hemodynamics and Sonomicrometry. Marielle Scherrer-Crosbie. 111, 2611-2616. 2005.

Isobe,M., Haber,E., and Khaw,B.A. (1991). Early detection of rejection and assessment of cyclosporine therapy by 111In antimyosin imaging in mouse heart allografts. Circulation. *84*, 1246-1255.

Isobe,M. and Ihara,A. (1993). Tolerance induction against cardiac allograft by anti-ICAM-1 and anti-LFA-1 treatment: T cells respond to in vitro allostimulation. Transplant Proc. *25*, 1079-1080.

Isobe,M., Kosuge,H., Koga,N., Futamatsu,H., and Suzuki,J. (2004). Gene therapy for heart transplantation-associated acute rejection, ischemia/reperfusion injury and coronary arteriosclerosis. Curr Gene Ther. *4*, 145-152.

Isobe,M., Narula,J., Southern,J.F., Strauss,H.W., Khaw,B.A., and Haber,E. (1992). Imaging the rejecting heart. In vivo detection of major histocompatibility complex class II antigen induction. Circulation. *85*, 738-746.

Isobe,M., Ohtani,H., Yagita,H., Okumura,K., Strauss,H.W., and Yazaki,Y. (1993). Detection of cardiac rejection in mice by radioimmune scintigraphy using 123iodine-labeled anti-ICAM-1 monoclonal antibody. Acta Cardiol. *48*, 235-243.

Janssen,B., Debets,J., Leenders,P., and Smits,J. (2002). Chronic measurement of cardiac output in conscious mice. Am J Physiol Regul. Integr. Comp Physiol. *282*, R928-R935.

Jassem,W., Fuggle,S.V., Rela,M., Koo,D.D., and Heaton,N.D. (2002). The role of mitochondria in ischemia/reperfusion injury. Transplantation. *73*, 493-499.

Jones,N.D., Turvey,S.E., Van Maurik,A., Hara,M., Kingsley,C.I., Smith,C.H., Mellor,A.L., Morris,P.J., and Wood,K.J. (2001). Differential Susceptibility of Heart, Skin, and Islet Allografts to T Cell-Mediated Rejection. The Journal of Immunology *166*, 2824-2830.

Ju,S.T., Cui,H., Panka,D.J., Ettinger,R., and Marshak-Rothstein,A. (1994). Participation of target Fas protein in apoptosis pathway induced by CD4+ Th1 and CD8+ cytotoxic T cells. Proc. Natl. Acad. Sci. U. S. A. *91*, 4185-4189.

Juncos,J.P., Grande,J.P., Murali,N., Croatt,A.J., Juncos,L.A., Hebbel,R.P., Katusic,Z.S., and Nath,K.A. (2006). Anomalous renal effects of tin protoporphyrin in a murine model of sickle cell disease. Am J Pathol. *169*, 21-31.

Kanno,S., Wu,Y.J., Lee,P.C., Dodd,S.J., Williams,M., Griffith,B.P., and Ho,C. (2001). Macrophage accumulation associated with rat cardiac allograft rejection detected by magnetic resonance imaging with ultrasmall superparamagnetic iron oxide particles. Circulation. *104*, 934-938.

Keidel WD (1950). Uber eine methode zur registrierung der Volumanderungen des Herzens am Menschen. Z Kreislaufforsch *39*, 257.

Kersten,J.R., Lowe,D., Hettrick,D.A., Pagel,P.S., Gross,G.J., and Warltier,D.C. (1996). Glyburide, a KATP channel antagonist, attenuates the cardioprotective effects of isoflurane in stunned myocardium. Anesth. Analg. *83*, 27-33.

Kessler,L.W., Korpel,A., and Palermo,P.R. (1972). Simultaneous acoustic and optical microscopy of biological specimens. Nature. *239*, 111-112.

Kevin Wei, Sanjiv Kaul. The coronary microcirculation in health and disease. Cardiology Clinics 22[2]. 2004. Cardiology Clinics.

Kimball,T.R., Semler,D.C., Witt,S.A., Khoury,P.R., and Daniels,S.R. (1997). Noninvasive markers for acute heart transplant rejection in children with the use of automatic border detection. J Am Soc. Echocardiogr. *10*, 964-972.

Klabunde,R.E. (2004). **Cardiovascular Physiology Concepts.** Lippincott Williams & Wilkins).

Klauss,V., Ackermann,K., Henneke,K.H., Spes,C., Zeitlmann,T., Werner,F., Regar,E., Rieber,J., Uberfuhr,P., Reichart,B., Theisen,K., and Mudra,H. (1997). Epicardial intimal thickening in transplant coronary artery disease and resistance vessel response to adenosine: a combined intravascular ultrasound and Doppler study. Circulation. *96*, II-64.

Klein,I., Samarel,A.M., Welikson,R., and Hong,C. (1991). Heterotopic cardiac transplantation decreases the capacity for rat myocardial protein synthesis. Circ Res. *68*, 1100-1107.

Koehl,G.E., Andrassy,J., Guba,M., Richter,S., Kroemer,A., Scherer,M.N., Steinbauer,M., Graeb,C., Schlitt,H.J., Jauch,K.W., and Geissler,E.K. (2004). Rapamycin protects allografts from rejection while simultaneously attacking tumors in immunosuppressed mice. Transplantation *77*, 1319-1326.

Kolar,F., MacNaughton,C., Papousek,F., and Korecky,B. (1993). Systolic mechanical performance of heterotopically transplanted hearts in rats treated with cyclosporin. Cardiovasc. Res. *27*, 1244-1247.

Konertz,W., Semik,M., and Bernhard,A. (1985). Heart, lung and heart-lung transplantation in rats. J Heart Transplant. *4*, 426-430.

Korecky,B. and Masika,M. (1991). Direct effect of increased hemodynamic load on cardiac mass. Circ Res. *68*, 1174-1178.

Kosuge,H., Haraguchi,G., Koga,N., Maejima,Y., Suzuki,J., and Isobe,M. (2006). Pioglitazone prevents acute and chronic cardiac allograft rejection. Circulation *113*, 2613-2622.

Kosuge,H., Suzuki,J., Gotoh,R., Koga,N., Ito,H., Isobe,M., Inobe,M., and Uede,T. (2003). Induction of immunologic tolerance to cardiac allograft by simultaneous blockade of inducible co-stimulator and cytotoxic T-lymphocyte antigen 4 pathway. Transplantation. *75*, 1374-1379.

Kreisel,D., Krupnick,A.S., Gelman,A.E., Engels,F.H., Popma,S.H., Krasinskas,A.M., Balsara,K.R., Szeto,W.Y., Turka,L.A., and Rosengard,B.R. (2002). Non-hematopoietic allograft cells directly activate CD8+ T cells and trigger acute rejection: an alternative mechanism of allorecognition. Nat. Med. *8*, 233-239.

Krieger,N.R. and Fathman,C.G. (1997). The use of CD4 and CD8 knockout mice to study the role of T-cell subsets in allotransplant rejection. J Heart Lung Transplant. *16*, 263-267.

Kulandavelu,S., Qu,D., and Adamson,S.L. (2006). Cardiovascular function in mice during normal pregnancy and in the absence of endothelial NO synthase. Hypertension. *47*, 1175-1182.

Larsen,C.P., Steinman,R.M., Witmer-Pack,M., Hankins,D.F., Morris,P.J., and Austyn,J.M. (1990). Migration and maturation of Langerhans cells in skin transplants and explants. J Exp. Med. *172*, 1483-1493.

Larsen,R.L., Applegate,P.M., Dyar,D.A., Ribeiro,P.A., Fritzsche,S.D., Mulla,N.F., Shirali,G.S., Kuhn,M.A., Chinnock,R.E., and Shah,P.M. (1998). Dobutamine stress echocardiography for assessing coronary artery disease after transplantation in children. J Am Coll Cardiol. *32*, 515-520.

Lemons,R.A. and Quate C.F. (1974). Acoustic microscopy: scanning version. Appl Phys Lett *4*, 163-165.

Libersan,D., Marchand,R., Montplaisir,S., Chartrand,C., and Dumont,L. (1997). Cardioprotective effects of diltiazem during acute rejection on heterotopic heart transplants. Eur. Surg. Res *29*, 229-236.

Lien,Y.C., Noel,T., Liu,H., Stromberg,A.J., Chen,K.C., and St.Clair,D.K. (2006). Phospholipase C-{delta}1 Is a Critical Target for Tumor Necrosis Factor Receptor-Mediated Protection against Adriamycin-Induced Cardiac Injury. Cancer Res *66*, 4329-4338.

Lindner,J.R. (2004). Microbubbles in medical imaging: current applications and future directions. Nat. Rev. Drug Discov. *3*, 527-532.

Lindner,J.R., Song,J., Xu,F., Klibanov,A.L., Singbartl,K., Ley,K., and Kaul,S. (2000). Noninvasive ultrasound imaging of inflammation using microbubbles targeted to activated leukocytes. Circulation. *102*, 2745-2750.

Lindner,V., Fingerle,J., and Reidy,M.A. (1993). Mouse model of arterial injury. Circ Res. *73*, 792-796.

Lorenz,J.N. and Robbins,J. (1997). Measurement of intraventricular pressure and cardiac performance in the intact closed-chest anesthetized mouse. Am J Physiol. *272*, H1137-H1146.

Lowry,R.P., Gurley,K.E., and Forbes,R.D. (1983). Immune mechanisms in organ allograft rejection. I. Delayed-type hypersensitivity and lymphocytotoxicity in heart graft rejection. Transplantation. *36*, 391-401.

Madamanchi,A. (2004). Mouse Models in Cardiology Research. McGill Journal of Medicine *8*, 34-39.

Makowka,L., Zerbe,T.R., Chapman,F., Qian,S.G., Sun,H., Murase,N., Kormos,R., Snyder,J., and Starzl,T.E. (1989). Prolonged rat cardiac preservation with UW lactobionate solution. Transplant Proc. *21*, 1350-1352.

Mannaerts,H.F., Balk,A.H., Simoons,M.L., Tijssen,J., van der Borden,S.G., Zondervan,P., Sutherland,G.R., and Roelandt,J.R. (1992). Changes in left ventricular function and wall thickness in heart transplant recipients and their relation to acute rejection: an assessment by digitised M mode echocardiography. Br. Heart J *68*, 356-364.

McGowan,B.S., Scott,C.B., Mu,A., McCormick,R.J., Thomas,D.P., and Margulies,K.B. (2003). Unloading-induced remodeling in the normal and hypertrophic left ventricle. AJP - Heart and Circulatory Physiology *284*, H2061-H2068.

McGregor,C.G., Hannan,J., Smith,A.F., Muir,A.L., and Wheatley,D.J. (1983). A study of cold cardioplegic myocardial protection in rats: an experimental model using the uptake of technetium 99m pyrophosphate and enzyme activity as parameters of injury. Cardiovasc Res. *17*, 70-74.

McGregor,C.G., McCallum,H.M., Hannan,J., Smith,A.F., and Muir,A.L. (1984). Long-term effects of cold cardioplegic myocardial protection in the rat. J Thorac Cardiovasc Surg. *87*, 913-919.

Meadows,W.R., Vanpraagh,S., Indreika,M., and Sharp,J.T. (1963). Premature mitral valve closure: a hemodynamic explanation for absence of the first sound in aortic insufficiency. Circulation. *28:251-8.*, 251-258.

Medawar,P.B. (1944). The behaviour and fate of skin autografts and skin homografts in rabbits: A report to the War Wounds Committee of the Medical Research Council. J Anat. *78*, 176-199.

Metzler,B., Mair,J., Lercher,A., Schaber,C., Hintringer,F., Pachinger,O., and Xu,Q. (2001). Mouse model of myocardial remodelling after ischemia: role of intercellular adhesion molecule-1. Cardiovasc Res. *49*, 399-407.

Meyer,C.R., Chiang,E.H., Fechner,K.P., Fitting,D.W., Williams,D.M., and Buda,A.J. (1988). Feasibility of high-resolution, intravascular ultrasonic imaging catheters. Radiology. *168*, 113-116.

Mintz,G.S., Popma,J.J., Pichard,A.D., Kent,K.M., Salter,L.F., Chuang,Y.C., Griffin,J., and Leon,M.B. (1996). Intravascular ultrasound predictors of restenosis after percutaneous transcatheter coronary revascularization. J Am Coll Cardiol. *27*, 1678-1687.

139

Mitsuhashi,N., Fischer-Lougheed,J., Shulkin,I., Kleihauer,A., Kohn,D.B., Weinberg,K.I., Starnes,V.A., and Kearns-Jonker,M. (2006). Tolerance induction by lentiviral gene therapy with a nonmyeloablative regimen. Blood. *107*, 2286-2293.

Miura,M., El Sawy,T., and Fairchild,R.L. (2003). Neutrophils Mediate Parenchymal Tissue Necrosis and Accelerate the Rejection of Complete Major Histocompatibility Complex-Disparate Cardiac Allografts in the Absence of Interferon-{gamma}. American Journal of Pathology *162*, 509-519.

Moidl,R., Chevtchik,O., Simon,P., Grimm,M., Wieselthaler,G., Ullrich,R., Mittlbock,M., Wolner,E., and Laufer,G. (1999). Noninvasive monitoring of peak filling rate with acoustic quantification echocardiography accurately detects acute cardiac allograft rejection. J Heart Lung Transplant. *18*, 194-201.

Moien-Afshari,F., McManus,B.M., and Laher,I. (2003). Immunosuppression and transplant vascular disease: benefits and adverse effects. Pharmacol. Ther. *100*, 141-156.

Morris,R.E. (1993). Commentary on new xenobiotic immunosuppressants for transplantation: Where are we, how did we get here, and where are we going? Clin. Transplant. *7*, 138-146.

Mottram,P.L., Pietersz,G.A., Smyth,M.J., Purcell,L.J., Clunie,G.J., and McKenzie,I.F. (1993). Evidence that an anthracycline-anti-CD8 immunoconjugate, idarubicin-anti-Ly-2.1, prolongs heart allograft survival in mice. Transplantation *55*, 484-490.

Mottram,P.L., Smith,J.A., Mason,A., Mirisklavos,A., Dumble,L.J., and Clunie,G.J. (1988). Electrocardiographic monitoring of cardiac transplants in mice. Cardiovasc. Res. *22*, 315-321.

Nakamura,A., Rokosh,D.G., Paccanaro,M., Yee,R.R., Simpson,P.C., Grossman,W., and Foster,E. (2001). LV systolic performance improves with development of hypertrophy after transverse aortic constriction in mice. AJP - Heart and Circulatory Physiology *281*, H1104-H1112.

Nakhleh,R.E., Jones,J., Goswitz,J.J., Anderson,E.A., and Titus,J. (1992). Correlation of endomyocardial biopsy findings with autopsy findings in human cardiac allografts. J Heart Lung Transplant. *11*, 479-485.

Nguyen,T.K., Nilakantan,V., Felix,C.C., Khanna,A.K., and Pieper,G.M. (2006). Beneficial effect of alpha-tocopheryl succinate in rat cardiac transplants. J Heart Lung Transplant. *25*, 707-715.

140

Niimi,M. (2001). The technique for heterotopic cardiac transplantation in mice: experience of 3000 operations by one surgeon. J Heart Lung Transplant. *20*, 1123-1128.

Nikolova,Z., Hof,A., Baumlin,Y., and Hof,R.P. (2001). Combined FTY720/cyclosporine treatment promotes graft survival and lowers the peripheral lymphocyte count in a murine cardiac allotransplantation model. Transplantation. *72*, 168-171.

Nissen,S.E., Grines,C.L., Gurley,J.C., Sublett,K., Haynie,D., Diaz,C., Booth,D.C., and DeMaria,A.N. (1990). Application of a new phased-array ultrasound imaging catheter in the assessment of vascular dimensions. In vivo comparison to cineangiography. Circulation. *81*, 660-666.

Nitenberg,A., Tavolaro,O., Loisance,D., Foult,J.M., Benhaiem,N., and Cachera,J.P. (1989). Severe impairment of coronary reserve during rejection in patients with orthotopic heart transplant. Circulation *79*, 59-65.

Nomura,T., Shibahara,T., Katakura,A., Matsubara,S., and Takano,N. (2006). Establishment of a murine model of bone invasion by oral squamous cell carcinoma. Oral Oncol. ..

O'Brien,P.D., O'Brien,W.D., Jr., Rhyne,T.L., Warltier,D.C., and Sagar,K.B. (1995). Relation of ultrasonic backscatter and acoustic propagation properties to myofibrillar length and myocardial thickness. Circulation *91*, 171-175.

Oei,F.B., Welters,M.J., Vaessen,L.M., Marquet,R.L., Zondervan,P.E., Weimar,W., and Bogers,A.J. (2000). Heart valve dysfunction resulting from cellular rejection in a novel heterotopic transplantation rat model. Transpl. Int. *13 Suppl 1*, S528-S531.

Ogawa,H., Mohiuddin,M.M., Yin,D.P., Shen,J., Chong,A.S., and Galili,U. (2004). Mouse-heart grafts expressing an incompatible carbohydrate antigen. II. Transition from accommodation to tolerance. Transplantation. *77*, 366-373.

Ohtani,H., Strauss,H.W., Southern,J.F., Tamatani,T., Miyasaka,M., Sekiguchi,M., and Isobe,M. (1995). Intercellular adhesion molecule-1 induction: a sensitive and quantitative marker for cardiac allograft rejection. J Am Coll Cardiol. *26*, 793-799.

Oluwole,S.F., Oluwole,O.O., DePaz,H.A., Adeyeri,A.O., Witkowski,P., and Hardy,M.A. (2003). CD4+CD25+ regulatory T cells mediate acquired transplant tolerance. Transpl. Immunol *11*, 287-293.

Ono,K. and Lindsey,E.S. (1969). Improved technique of heart transplantation in rats. J Thorac Cardiovasc Surg. *57*, 225-229.

Pahl,E., Crawford,S.E., Swenson,J.M., Duffy,C.E., Fricker,F.J., Backer,C.L., Mavroudis,C., and Chaudhry,F.A. (1999). Dobutamine stress echocardiography: experience in pediatric heart transplant recipients. J Heart Lung Transplant. *18*, 725-732.

Paulsen,W., Magid,N., Sagar,K., Hastillo,A., Wolfgang,T.C., Lower,R.R., and Hess,M.L. (1985). Left ventricular function of heart allografts during acute rejection: an echocardiographic assessment. J Heart Transplant *4*, 525-529.

Pavlin CJ and Foster FS (1995). Ultrasound biomicroscopy of the eye. (New York: Springer-Verlag).

Pavlin,C.J., Sherar,M.D., and Foster,F.S. (1990). Subsurface ultrasound microscopic imaging of the intact eye. Ophthalmology. *97*, 244-250.

Pearson,T.C., Alexander,D.Z., Winn,K.J., Linsley,P.S., Lowry,R.P., and Larsen,C.P. (1994). Transplantation tolerance induced by CTLA4-Ig. Transplantation. *57*, 1701-1706.

Penno,E., Johnsson,C., Johansson,L., and Ahlstrom,H. (2006). Macrophage uptake of ultra-small iron oxide particles for magnetic resonance imaging in experimental acute cardiac transplant rejection. Acta Radiol. *47*, 264-271.

Peteiro,J., Redondo,F., Calvino,R., Cuenca,J., Pradas,G., and Castro,B.A. (1996). Differences in heart transplant physiology according to surgical technique. J Thorac Cardiovasc Surg. *112*, 584-589.

Petri,W.A., Jr. (1994). Infections in heart transplant recipients. Clin. Infect. Dis. *18*, 141-146.

Phoon,C.K., Ji,R.P., Aristizabal,O., Worrad,D.M., Zhou,B., Baldwin,H.S., and Turnbull,D.H. (2004). Embryonic heart failure in NFATc1-/- mice: novel mechanistic insights from in utero ultrasound biomicroscopy. Circ Res. *95*, 92-99.

Phoon,C.K. and Turnbull,D.H. (2003). Ultrasound biomicroscopy-Doppler in mouse cardiovascular development. Physiol Genomics. *14*, 3-15.

Pietra,B.A. and Gill,R.G. (2001). Immunobiology of cardiac allograft and xenograft transplantation. Semin. Thorac Cardiovasc Surg Pediatr. Card Surg Annu. *4:123-57.*, 123-157.

Planas,R., Alba,A., Carrillo,J., Puertas,M.C., Ampudia,R., Pastor,X., Okamoto,H., Takasawa,S., Gurr,W., Pujol-Borrell,R., Verdaguer,J., and Vives-Pi,M. (2006). Reg (regenerating) gene overexpression in islets from non-obese diabetic mice with accelerated diabetes: role of IFNbeta. Diabetologia. ..

142

Pollick,C., Hale,S.L., and Kloner,R.A. (1995). Echocardiographic and cardiac Doppler assessment of mice. J Am Soc. Echocardiogr. *8*, 602-610.

Puleo,J.A., Aranda,J.M., Weston,M.W., Cintron,G., French,M., Clark,L., and Fontanet,H.L. (1998). Noninvasive detection of allograft rejection in heart transplant recipients by use of Doppler tissue imaging. J Heart Lung Transplant. *17*, 176-184.

Quinones,M.A., Otto,C.M., Stoddard,M., Waggoner,A., and Zoghbi,W.A. (2002). Recommendations for quantification of Doppler echocardiography: a report from the Doppler Quantification Task Force of the Nomenclature and Standards Committee of the American Society of Echocardiography. J Am Soc. Echocardiogr. *15*, 167-184.

Raisky,O., Gomez,L., Chalabreysse,L., Gateau-Roesch,O., Loufouat,J., Thivolet-Bejui,F., Ninet,J., and Ovize,M. (2004). Mitochondrial permeability transition in cardiomyocyte apoptosis during acute graft rejection. Am J Transplant *4*, 1071-1078.

Rakusan,K., Heron,M.I., Kolar,F., and Korecky,B. (1997). Transplantation-induced atrophy of normal and hypertrophic rat hearts: effect on cardiac myocytes and capillaries. J Mol. Cell Cardiol. *29*, 1045-1054.

Riess,M.L., Stowe,D.F., and Warltier,D.C. (2004). Cardiac pharmacological preconditioning with volatile anesthetics: from bench to bedside? AJP - Heart and Circulatory Physiology *286*, H1603-H1607.

Rodrigues,A.C., Frimm,C.C., Bacal,F., Andreolli,V., Tsutsui,J.M., Bocchi,E.A., Mathias,W., Jr., and Lage,S.G. (2005). Coronary flow reserve impairment predicts cardiac events in heart transplant patients with preserved left ventricular function. Int. J Cardiol *103*, 201-206.

Ross,H.J., Gullestad,L., Hunt,S.A., Tovey,D.A., Puryear,J.B., McMillan,A., Stinson,E.B., and Valantine,H.A. (1996). Early Doppler echocardiographic dysfunction is associated with an increased mortality after orthotopic cardiac transplantation. Circulation. *94*, II289-II293.

Roth,D.M., Swaney,J.S., Dalton,N.D., Gilpin,E.A., and Ross,J., Jr. (2002). Impact of anesthesia on cardiac function during echocardiography in mice. Am J Physiol Heart Circ Physiol. *282*, H2134-H2140.

Sade,L.E., Sezgin,A., Ulucam,M., Taymaz,S., Simsek,V., Tayfun,E., Tokel,K., Aslamaci,S., and Muderrisoglu,H. (2006). Evaluation of the potential role of echocardiography in the detection of allograft rejection in heart transplant recipients. Transplant Proc. *38*, 636-638.

Sagar,K.B., Hastillo,A., Wolfgang,T.C., Lower,R.R., and Hess,M.L. (1981). Left ventricular mass by M-mode echocardiography in cardiac transplant patients with acute rejection. Circulation *64*, II217-II220.

Sahn,D.J., DeMaria,A., Kisslo,J., and Weyman,A. (1978). Recommendations regarding quantitation in M-mode echocardiography: results of a survey of echocardiographic measurements. Circulation. *58*, 1072-1083.

Sands,K.E., Appel,M.L., Lilly,L.S., Schoen,F.J., Mudge,G.H., Jr., and Cohen,R.J. (1989). Power spectrum analysis of heart rate variability in human cardiac transplant recipients. Circulation. *79*, 76-82.

Scarabelli,T.M., Stephanou,A., Pasini,E., Comini,L., Raddino,R., Knight,R.A., and Latchman,D.S. (2002). Different Signaling Pathways Induce Apoptosis in Endothelial Cells and Cardiac Myocytes During Ischemia/Reperfusion Injury. Circ Res *90*, 745-748.

Schaefer,A., Klein,G., Brand,B., Lippolt,P., Drexler,H., and Meyer,G.P. (2003). Evaluation of left ventricular diastolic function by pulsed Doppler tissue imaging in mice. J Am Soc. Echocardiogr. *16*, 1144-1149.

Scherrer-Crosbie,M. (2006). Role of echocardiography in studies of murine models of cardiac diseases. Arch. Mal Coeur Vaiss. *99*, 237-241.

Scherrer-Crosbie,M., Glysing-Jensen,T., Fry,S.J., Vancon,A.C., Gadiraju,S., Picard,M.H., and Russell,M.E. (2002). Echocardiography improves detection of rejection after heterotopic mouse cardiac transplantation. J Am Soc. Echocardiogr. *15*, 1315-1320.

Scherrer-Crosbie,M., Steudel,W., Hunziker,P.R., Foster,G.P., Garrido,L., Liel-Cohen,N., Zapol,W.M., and Picard,M.H. (1998). Determination of right ventricular structure and function in normoxic and hypoxic mice: a transesophageal echocardiographic study. Circulation. *98*, 1015-1021.

Scherrer-Crosbie,M., Steudel,W., Hunziker,P.R., Liel-Cohen,N., Ullrich,R., Zapol,W.M., and Picard,M.H. (1999). Three-dimensional echocardiographic assessment of left ventricular wall motion abnormalities in mouse myocardial infarction. J Am Soc. Echocardiogr. *12*, 834-840.

Schiller,N.B., Shah,P.M., Crawford,M., DeMaria,A., Devereux,R., Feigenbaum,H., Gutgesell,H., Reichek,N., Sahn,D., Schnittger,I., and . (1989). Recommendations for quantitation of the left ventricle by two-dimensional echocardiography. American Society of Echocardiography Committee on Standards, Subcommittee on Quantitation of Two-Dimensional Echocardiograms. J Am Soc. Echocardiogr. *2*, 358-367.

Sebag,I.A., Handschumacher,M.D., Ichinose,F., Morgan,J.G., Hataishi,R., Rodrigues,A.C., Guerrero,J.L., Steudel,W., Raher,M.J., Halpern,E.F., Derumeaux,G., Bloch,K.D., Picard,M.H., and Scherrer-Crosbie,M. (2005). Quantitative assessment of regional myocardial function in mice by tissue Doppler imaging: comparison with hemodynamics and sonomicrometry. Circulation. *111*, 2611-2616.

Shepler,S.A. and Patel,A.N. (2007). Cardiac cell therapy: a treatment option for cardiomyopathy. Crit Care Nurs. Q. *30*, 74-80.

Sherar,M.D. and Foster,F.S. (1989). The design and fabrication of high frequency poly(vinylidene fluoride) transducers. Ultrason. Imaging. *11*, 75-94.

Sherar,M.D., Starkoski,B.G., Taylor,W.B., and Foster,F.S. (1989). A 100 MHz B-scan ultrasound backscatter microscope. Ultrason. Imaging. *11*, 95-105.

Shimmura,H., Tanabe,K., Habiro,K., Abe,R., and Toma,H. (2006). Combination effect of mycophenolate mofetil with mizoribine on cell proliferation assays and in a mouse heart transplantation model. Transplantation. *82*, 175-179.

Silber,S. (1979). The Rat in Microsurgery. (Baltimore, MD, USA: Williams & Wilkins).

Smith,J.A., Mottram,P.L., Mirisklavos,A., Mason,A., Dumble,L.J., and Clunie,G.J. (1987). The effect of operative ischemia in murine cardiac transplantation: isograft control studies. Surgery *101*, 86-90.

Sokolov,S.J. (1935). Ultrasonic oscillations and their applications. Tech Phys USSR *2*, 522-534.

Soto,P.F., Jia,C.X., Carter,Y.M., Rabkin,D., Starr,J.P., Amirhamzeh,M.M., Hsu,D.T., Sciacca,R., Fisher,P.E., and Spotnitz,H.M. (1998). Effect of improved myocardial protection on edema and diastolic properties of the rat left ventricle during acute allograft rejection. J Heart Lung Transplant. *17*, 608-616.

Spencer,A.U., Hart,J.P., Cabreriza,S.E., Rabkin,D.G., Weinberg,A.D., and Spotnitz,H.M. (2003). Aortic regurgitation in the heterotopic rat heart transplant: effect on ventricular remodeling and diastolic function. J Heart Lung Transplant. *22*, 937-945.

Spriet,M.P., Girard,C.A., Foster,S.F., Harasiewicz,K., Holdsworth,D.W., and Laverty,S. (2005). Validation of a 40 MHz B-scan ultrasound biomicroscope for the evaluation of osteoarthritis lesions in an animal model. Osteoarthritis. Cartilage. *13*, 171-179.

St Goar,F.G., Pinto,F.J., Alderman,E.L., Valantine,H.A., Schroeder,J.S., Gao,S.Z., Stinson,E.B., and Popp,R.L. (1992). Intracoronary ultrasound in cardiac transplant recipients. In vivo evidence of "angiographically silent" intimal thickening. Circulation. *85*, 979-987.

Steinbruchel,D.A., Madsen,H.H., Nielsen,B., Kemp,E., Larsen,S., and Koch,C. (1991). The effect of combined treatment with total lymphoid irradiation, cyclosporin A, and anti-CD4 monoclonal antibodies in a hamster-to-rat heart transplantation model. Transplant Proc. *23*, 579-580.

Steinmuller,D. (2001). Skin allograft rejection by stable hematopoietic chimeras that accept organ allografts sill is an enigma. Transplantation. *72*, 8-9.

Steinmuller,D., Wakely,E., and Landas,S.K. (1991). Evidence that epidermal alloantigen Epa-1 is an immunogen for murine heart as well as skin allograft rejection. Transplantation. *51*, 459-463.

Stempfle,H.U., Angermann,C.E., Kraml,P., Schutz,A., Kemkes,B.M., and Theisen,K. (1993). Serial changes during acute cardiac allograft rejection: quantitative ultrasound tissue analysis versus myocardial histologic findings. J Am Coll Cardiol. *22*, 310-317.

Stewart,S., Winters,G.L., Fishbein,M.C., Tazelaar,H.D., Kobashigawa,J., Abrams,J., Andersen,C.B., Angelini,A., Berry,G.J., Burke,M.M., Demetris,A.J., Hammond,E., Itescu,S., Marboe,C.C., McManus,B., Reed,E.F., Reinsmoen,N.L., Rodriguez,E.R., Rose,A.G., Rose,M., Suciu-Focia,N., Zeevi,A., and Billingham,M.E. (2005). Revision of the 1990 working formulation for the standardization of nomenclature in the diagnosis of heart rejection. J Heart Lung Transplant. *24*, 1710-1720.

Sullivan,P.M., Mezdour,H., Quarfordt,S.H., and Maeda,N. (1998). Type III hyperlipoproteinemia and spontaneous atherosclerosis in mice resulting from gene replacement of mouse Apoe with human Apoe*2. J Clin. Invest. *102*, 130-135.

Sun,J., Sheil,A.G., Wang,C., Wang,L., Rokahr,K., Sharland,A., Jung,S.E., Li,L., McCaughan,G.W., and Bishop,G.A. (1996). Tolerance to rat liver allografts: IV. Acceptance depends on the quantity of donor tissue and on donor leukocytes. Transplantation. *62*, 1725-1730.

Superina,R.A., Peugh,W.N., Wood,K.J., and Morris,P.J. (1986). Assessment of primarily vascularized cardiac allografts in mice. Transplantation *42*, 226-227.

Suzuki,J., Isobe,M., Izawa,A., Takahashi,W., Yamazaki,S., Okubo,Y., Amano,J., and Sekiguchi,M. (1999). Differential Th1 and Th2 cell regulation of murine cardiac allograft acceptance by blocking cell adhesion of ICAM-1/LFA-1 and VCAM-1/VLA-4. Transpl. Immunol. *7*, 65-72.

Suzuki,J., Isobe,M., Yamazaki,S., Horie,S., Okubo,Y., and Sekiguchi,M. (1998). Sensitive diagnosis of cardiac allograft rejection by detection of cytokine transcription in situ. Cardiovasc Res. *40*, 307-313.

Szabo,G., Batkai,S., Dengler,T.J., Bahrle,S., Stumpf,N., Notmeyer,W., Zimmermann,R., Vahl,C.F., and Hagl,S. (2001). Systolic and diastolic properties and myocardial blood flow in the heterotopically transplanted rat heart during acute cardiac rejection. World J Surg. *25*, 545-552.

Tanaka,M., Gunawan,F., Terry,R.D., Inagaki,K., Caffarelli,A.D., Hoyt,G., Tsao,P.S., Mochly-Rosen,D., and Robbins,R.C. (2005a). Inhibition of heart transplant injury and graft coronary artery disease after prolonged organ ischemia by selective protein kinase C regulators. J Thorac. Cardiovasc. Surg. *129*, 1160-1167.

Tanaka,M., Mokhtari,G.K., Balsam,L.B., Cooke,D.T., Kofidis,T., Zwierzchonievska,M., and Robbins,R.C. (2005b). Cyclosporine mitigates graft coronary artery disease in murine cardiac allografts: description and validation of a novel fully allogeneic model. J Heart Lung Transplant. *24*, 446-453.

Tanaka,M., Mokhtari,G.K., Terry,R.D., Gunawan,F., Balsam,L.B., Hoyt,G., Lee,K.H., Tsao,P.S., and Robbins,R.C. (2005c). Prolonged cold ischemia in rat cardiac allografts promotes ischemia-reperfusion injury and the development of graft coronary artery disease in a linear fashion. J Heart Lung Transplant. *24*, 1906-1914.

Tanaka,M., Swijnenburg,R.J., Gunawan,F., Cao,Y.A., Yang,Y., Caffarelli,A.D., de Bruin,J.L., Contag,C.H., and Robbins,R.C. (2005d). In vivo visualization of cardiac allograft rejection and trafficking passenger leukocytes using bioluminescence imaging. Circulation. *112*, I105-I110.

Tei,C. (1995). New non-invasive index for combined systolic and diastolic ventricular function. J Cardiol. *26*, 135-136.

Turnbull,D.H., Bloomfield,T.S., Baldwin,H.S., Foster,F.S., and Joyner,A.L. (1995). Ultrasound backscatter microscope analysis of early mouse embryonic brain development. Proc. Natl. Acad. Sci. U. S. A. *92*, 2239-2243.

Turnbull,D.H., Ramsay,J.A., Shivji,G.S., Bloomfield,T.S., From,L., Sauder,D.N., and Foster,F.S. (1996). Ultrasound backscatter microscope analysis of mouse melanoma progression. Ultrasound Med. Biol. *22*, 845-853.

Valantine,H.A., Appleton,C.P., Hatle,L.K., Hunt,S.A., Billingham,M.E., Shumway,N.E., Stinson,E.B., and Popp,R.L. (1989). A hemodynamic and Doppler echocardiographic study of ventricular function in long-term cardiac allograft recipients. Etiology and prognosis of restrictive-constrictive physiology. Circulation *79*, 66-75.

147

Valantine,H.A., Fowler,M.B., Hunt,S.A., Naasz,C., Hatle,L.K., Billingham,M.E., Stinson,E.B., and Popp,R.L. (1987). Changes in Doppler echocardiographic indexes of left ventricular function as potential markers of acute cardiac rejection. Circulation. *76*, V86-V92.

van den Maagdenberg,A.M., Hofker,M.H., Krimpenfort,P.J., de,B., I, van Vlijmen,B., van der,B.H., Havekes,L.M., and Frants,R.R. (1993). Transgenic mice carrying the apolipoprotein E3-Leiden gene exhibit hyperlipoproteinemia. J Biol. Chem. *268*, 10540-10545.

Vandenberg,B.F., Mohanty,P.K., Craddock,K.J., Barnhart,G., Hanrahan,J., Szentpetery,S., and Lower,R.R. (1988). Clinical significance of pericardial effusion after heart transplantation. J Heart Transplant. *7*, 128-134.

Vivekananthan,K., Kalapura,T., Mehra,M., Lavie,C., Milani,R., Scott,R., and Park,M. (2002). Usefulness of the combined index of systolic and diastolic myocardial performance to identify cardiac allograft rejection. Am J Cardiol *90*, 517-520.

Wada,T., Ono,K., Hadama,T., Uchida,Y., Shimada,T., and Arita,M. (1999). Detection of acute cardiac rejection by analysis of heart rate variability in heterotopically transplanted rats. J Heart Lung Transplant. *18*, 499-509.

Wang,H., Hosiawa,K.A., Min,W., Yang,J., Zhang,X., Garcia,B., Ichim,T.E., Zhou,D., Lian,D., Kelvin,D.J., and Zhong,R. (2003). Cytokines regulate the pattern of rejection and susceptibility to cyclosporine therapy in different mouse recipient strains after cardiac allografting. J Immunol *171*, 3823-3836.

Weis,M. (2002). Cardiac allograft vasculopathy: prevention and treatment options. Transplant Proc. *34*, 1847-1849.

Weitzman,L.B., Tinker,W.P., Kronzon,I., Cohen,M.L., Glassman,E., and Spencer,F.C. (1984). The incidence and natural history of pericardial effusion after cardiac surgery-- an echocardiographic study. Circulation. *69*, 506-511.

Weller,G.E., Lu,E., Csikari,M.M., Klibanov,A.L., Fischer,D., Wagner,W.R., and Villanueva,F.S. (2003). Ultrasound imaging of acute cardiac transplant rejection with microbubbles targeted to intercellular adhesion molecule-1. Circulation. *108*, 218-224.

Welsh,D.C., Dipla,K., McNulty,P.H., Mu,A., Ojamaa,K.M., Klein,I., Houser,S.R., and Margulies,K.B. (2001a). Preserved contractile function despite atrophic remodeling in unloaded rat hearts. Am J Physiol Heart Circ Physiol. *281*, H1131-H1136.

Welsh,D.C., Dipla,K., McNulty,P.H., Mu,A., Ojamaa,K.M., Klein,I., Houser,S.R., and Margulies,K.B. (2001b). Preserved contractile function despite atrophic remodeling in unloaded rat hearts. AJP - Heart and Circulatory Physiology *281*, H1131-H1136.

West,L.J., Morris,P.J., and Wood,K.J. (1994). Neonatal induction of tolerance to cardiac allografts. Transplant Proc. *26*, 207-208.

West,L.J., Morris,P.J., and Wood,K.J. (1995). Neonatally induced transplantation tolerance to primarily vascularised cardiac allografts is not donor-specific. Transplant Proc. *27*, 184-185.

West,L.J., Pollock-Barziv,S.M., Dipchand,A.I., Lee,K.J., Cardella,C.J., Benson,L.N., Rebeyka,I.M., and Coles,J.G. (2001). ABO-incompatible heart transplantation in infants. N Engl J Med. *344*, 793-800.

West,L.J. and Tao,K. (2002). Acceptance of third-party cardiac but not skin allografts induced by neonatal exposure to semi-allogeneic lymphohematopoietic cells. Am J Transplant *2*, 733-744.

White,W.L., Zhang,Y.L., Shelby,J., Trautman,M.S., Perkins,S.L., Hammond,E.H., and Shaddy,R.E. (1997). Myocardial apoptosis in a heterotopic murine heart transplantation model of chronic rejection and graft vasculopathy. J Heart Lung Transplant *16*, 250-255.

Wikstrom,J., Gronros,J., Bergstrom,G., and Gan,L.M. (2005). Functional and morphologic imaging of coronary atherosclerosis in living mice using high-resolution color Doppler echocardiography and ultrasound biomicroscopy. J Am Coll Cardiol. *46*, 720-727.

Wild,J.J. and Neal,D. (1951). Use of high-frequency ultrasonic waves for detecting changes of texture in living tissues. Lancet. *1*, 655-657.

Wu,Y.J., Sato,K., Ye,Q., and Ho,C. (2004). MRI investigations of graft rejection following organ transplantation using rodent models. Methods Enzymol. *386:73-105.*, 73-105.

Wu,Y.L., Ye,Q., Foley,L.M., Hitchens,T.K., Sato,K., Williams,J.B., and Ho,C. (2006). In situ labeling of immune cells with iron oxide particles: an approach to detect organ rejection by cellular MRI. Proc. Natl. Acad. Sci. U. S. A. *103*, 1852-1857.

Yamada,A., Laufer,T.M., Gerth,A.J., Chase,C.M., Colvin,R.B., Russell,P.S., Sayegh,M.H., and Auchincloss,H., Jr. (2003). Further analysis of the T-cell subsets and pathways of murine cardiac allograft rejection. Am J Transplant. *3*, 23-27.

Yamaura,K., Ito,K., Tsukioka,K., Wada,Y., Makiuchi,A., Sakaguchi,M., Akashima,T., Fujimori,M., Sawa,Y., Morishita,R., Matsumoto,K., Nakamura,T., Suzuki,J., Amano,J., and Isobe,M. (2004). Suppression of acute and chronic rejection by hepatocyte growth factor in a murine model of cardiac transplantation: induction of tolerance and prevention of cardiac allograft vasculopathy. Circulation *110*, 1650-1657.

Yang,X.P., Liu,Y.H., Rhaleb,N.E., Kurihara,N., Kim,H.E., and Carretero,O.A. (1999). Echocardiographic assessment of cardiac function in conscious and anesthetized mice. AJP - Heart and Circulatory Physiology *277*, H1967-H1974.

Yock,P.G., Linker,D.T., and Angelsen,B.A. (1989). Two-dimensional intravascular ultrasound: technical development and initial clinical experience. J Am Soc. Echocardiogr. *2*, 296-304.

Yokode,M., Hammer,R.E., Ishibashi,S., Brown,M.S., and Goldstein,J.L. (1990). Diet-induced hypercholesterolemia in mice: prevention by overexpression of LDL receptors. Science. *250*, 1273-1275.

Yoshida,S., Takeuchi,K., del Nido,P.J., and Ho,C. (1998). Diastolic dysfunction coincides with early mild transplant rejection: in situ measurements in a heterotopic rat heart transplant model. J Heart Lung Transplant. *17*, 1049-1056.

Yoshida,T., Mori,M., and Nimura,Y. (1956). Study on examining the heart with the ultrasonic Doppler method. Jpn Circ J *20*.

Yuh,D.D. and Morris,R.E. (1993). The immunopharmacology of immunosuppression by 15-deoxyspergualin. Transplantation. *55*, 578-591.

Zbilut,J.P., Murdock,D.K., Lawson,L., Lawless,C.E., Von Dreele,M.M., and Porges,S.W. (1988). Use of power spectral analysis of respiratory sinus arrhythmia to detect graft rejection. J Heart Transplant. *7*, 280-288.

Zhang,S.H., Reddick,R.L., Piedrahita,J.A., and Maeda,N. (1992). Spontaneous hypercholesterolemia and arterial lesions in mice lacking apolipoprotein E. Science. *258*, 468-471.

Zhou, Y. Q., Bishay, R, Feintuch, A., Tao, K., Golding, F., Zhu, W, West, L. J., and Henkelman, R. M. (2007) Morphological and functional evaluation of murine heterotopic cardiac grafts using ultrasound biomicroscopy. Ultrasound Med. Biol. *33*, 870-9.

Zhou,Y.Q., Foster,F.S., Nieman,B.J., Davidson,L., Chen,X.J., and Henkelman,R.M. (2004). Comprehensive transthoracic cardiac imaging in mice using ultrasound biomicroscopy with anatomical confirmation by magnetic resonance imaging. Physiol. Genomics *18*, 232-244.

Zimmermann,W.H., Didie,M., Doker,S., Melnychenko,I., Naito,H., Rogge,C., Tiburcy,M., and Eschenhagen,T. (2006a). Heart muscle engineering: an update on cardiac muscle replacement therapy. Cardiovasc Res. *71*, 419-429.

Zimmermann,W.H., Melnychenko,I., Wasmeier,G., Didie,M., Naito,H., Nixdorff,U., Hess,A., Budinsky,L., Brune,K., Michaelis,B., Dhein,S., Schwoerer,A., Ehmke,H., and Eschenhagen,T. (2006b). Engineered heart tissue grafts improve systolic and diastolic function in infarcted rat hearts. Nat. Med. *12*, 452-458.

VI. APPENDIX

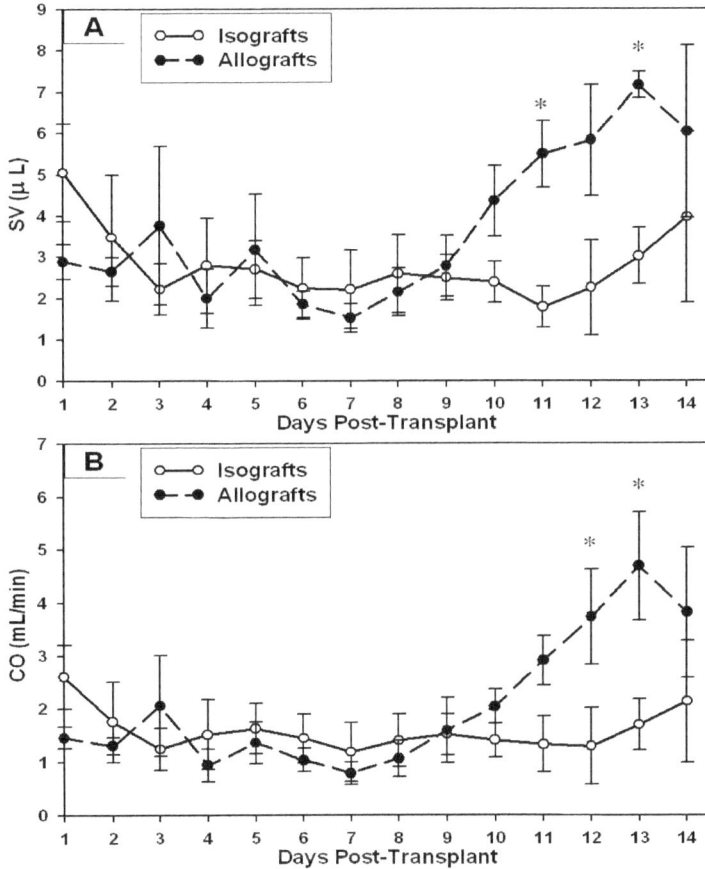

Figure A1 – Stroke Volume and Cardiac Output in Isografts and Allografts.
Automated LV functional analysis generated SV and CO values from semi-automated
LV endocardial traces. **A.** Statistical differences in stroke volume between isografts and
allografts are seen on days 11 and 13 only; SV in both groups increased late after
transplant. **B.** Cardiac output is similar for iso- and allografts early post-transplant. A
trend for allografts to increase late after transplant is seen. Differences are seen on days
12 and 13. The data was analyzed using two-way repeated-measures ANOVA followed
by post hoc testing by the Holm-Sidak method for all pair-wise comparisons. Data are
reported as means ± SEM. (*) indicates significance at p<0.05 between iso- and
allografts.

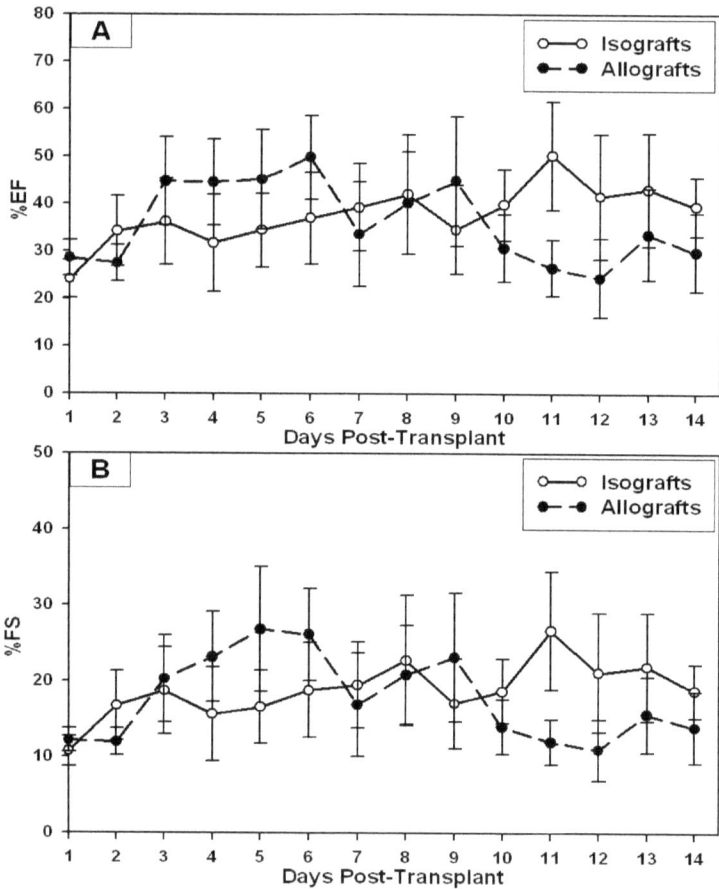

Figure A2 — %Ejection fraction (%EF) and %fractional shortening (%FS) in Isografts and Allografts. Automated LV functional analysis generated %EF and %FS values from semi-automated LV endocardial traces. **A.** %EF is variable for both iso- and allografts, with day 14 rejecting grafts reaching figures close to day 1. No statistical differences were detected between the two groups. **B.** %FS yields a similar temporal pattern as %EF for both groups of grafts. Like %EF, no statistical differences were seen between iso- and allografts. The data was analyzed using two-way repeated-measures ANOVA followed. No significance was detected at $p<0.05$ between iso- and allografts.

The temporal patterns for changes in CO were similar for both isografts and allografts for the majority of the post-transplant period. SV in both groups increased late after transplant. Isograft and allograft %EF and %FS also showed similar temporal patterns though no significant differences were detected between isografts or allografts.

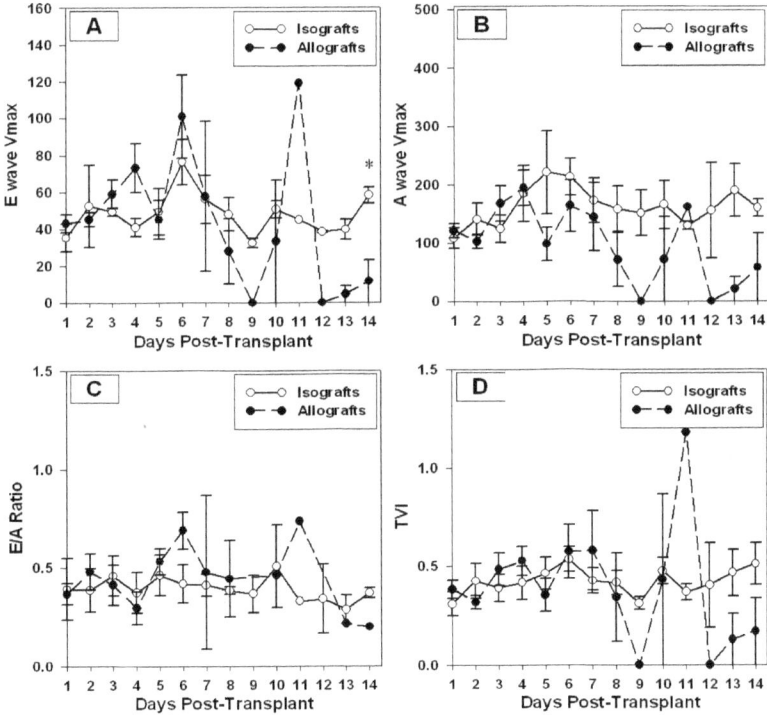

Figure A3 - Tricuspid Doppler Flow Measurements. A. Peak E-wave velocity of the biphasic tricuspid Doppler waveform. **B.** Maximum A-wave flow velocity. **C.** E/A ratios. **D.** The time-velocity integral of the tricuspid. Data are reported as mean $\pm$ SEM. (*) indicates significance at $P \leq 0.001$ between iso- and allografts.

Peak E-wave velocity of the biphasic tricuspid Doppler waveform was similar for both grafts from days 1-6 but variable afterwards, showing differences between iso- and allografts on day 14 only. Maximum A-wave flow velocity was also variable, though higher values seemed to be observed in isografts when compared to allografts after day 5. E/A ratios were variable for both isografts and allografts and no significant differences were detected in the post-transplant period. The time-velocity integral (TVI) of the tricuspid waveform was stable and similar for both groups of grafts from days 1-7 but became variable in allografts late in the post-transplant period. As with the majority of the tricuspid parameters analyzed, no significant differences were detected.

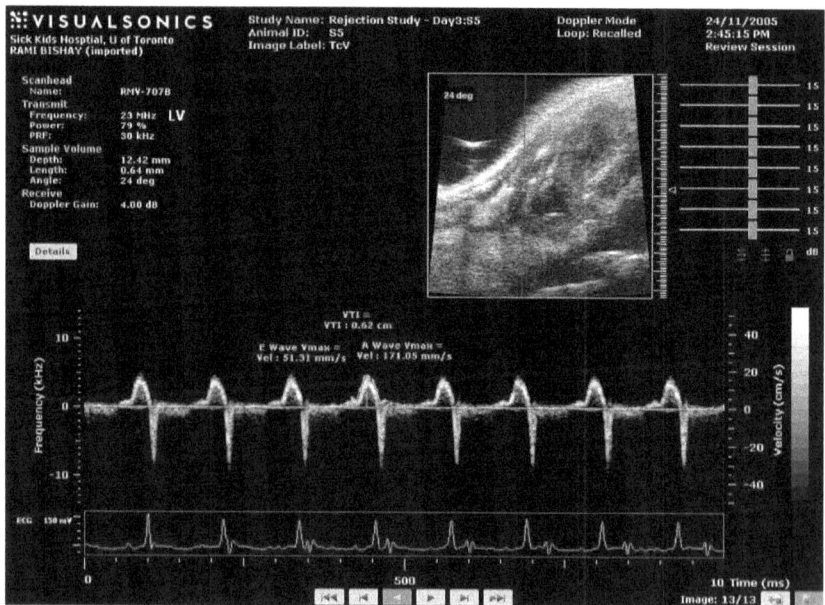

Figure A4 – Tricuspid Doppler Waveforms: Doppler diastolic inflow waveforms were obtained from a modified LAV.

Tricuspid Doppler inflow was easily acquired using UBM-Doppler from a modified LAV, however, the small, variable, and inconsistent patterns seen in the biphasic waveforms of tricuspid flow did not yield any significant differences between the groups of grafts or any temporal changes within each experimental group.

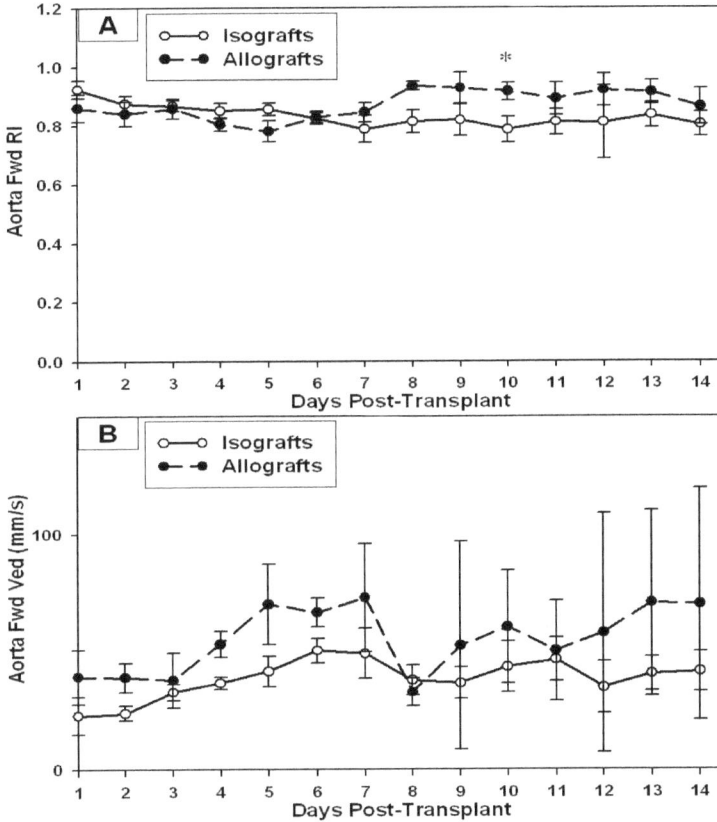

Figure A5 – Aortic Forward Resistance Index (RI) and End-Diastolic Velocity (Ved).
A. Fwd RI is similar for both isografts and allografts throughout the post-transplant period (0.8-1.0), with only transient statistical differences detected on day 10. **B.** Aortic Ved yields a similar temporal pattern for both groups of grafts, though allografts reach a higher peak on day 6. Isografts yield low Ved values (<50 mm/s) post-transplant; allografts reach values close to isografts late in rejection. No statistical differences between the two groups of grafts were detected. The data was analyzed using two-way repeated-measures ANOVA followed by post hoc testing by the Holm-Sidak method. Data are reported as means ± SEM. (*) indicates significance at p<0.05 between iso- and allografts.

Aortic forward RI was linearly constant and similar for both isografts and allografts throughout the post-transplant period (0.8-1.0). Aortic Ved yields a similar temporal pattern for both groups of grafts, though allografts reach a higher peak on day 6. Isografts yield low

Ved values (<50 mm/s) post-transplant whereas allografts reach values close to isografts late in rejection. No statistical differences between the two groups of grafts were detected.

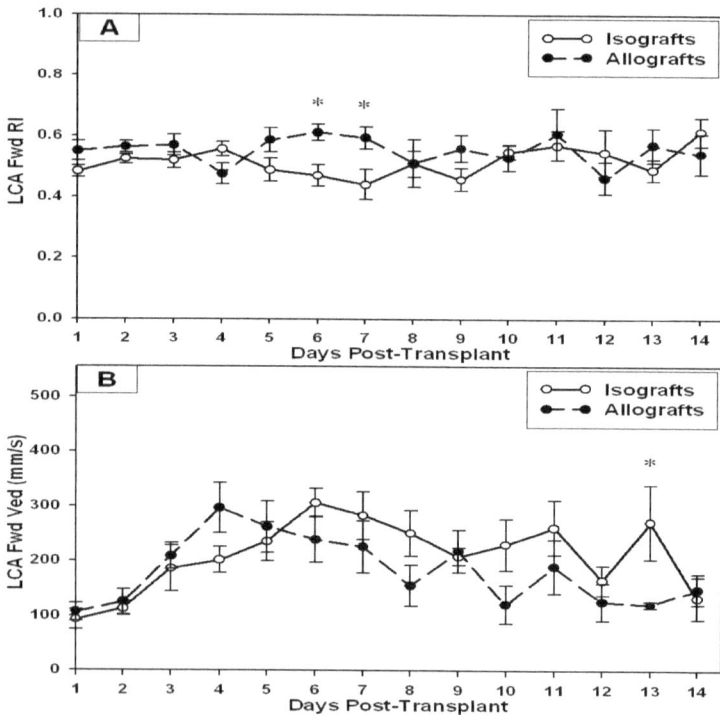

Figure A6 – Forward Resistance Index (RI) and End-Diastolic Velocity (Ved) in the Left Coronary Artery. A. A modified RI was calculated as the percent difference between Vmax and Ved divided by Vmax. Transient statistical differences were detected on days 6 and 7 only. RI values resided in the 0.4-0.6 range. B. Fwd Ved was determined as the lowest notch in the highest forward waveform. Ved yields a similar temporal pattern for both groups of grafts from days 1-3, with allografts peaking at day 4 and isografts at day 6. Allografts gradually decrease after day 4. Differences were detected on day 13 only. The data was analyzed using two-way repeated-measures ANOVA followed by post hoc testing by the Holm-Sidak method. Data are reported as means ± SEM. (*) indicates significance at p<0.05 between iso- and allografts.

No statistical differences were detected in LCA RI between isografts and allografts; RI values resided in the 0.4-0.6 range. LCA Fwd Ved yields a similar temporal pattern for both groups of grafts from days 1-3, with allografts peaking at day 4 and isografts at day 6.

Allografts gradually decrease after day 4. A transient statistical difference from allografts was detected on day 13 only ($*P \leq 0.001$).

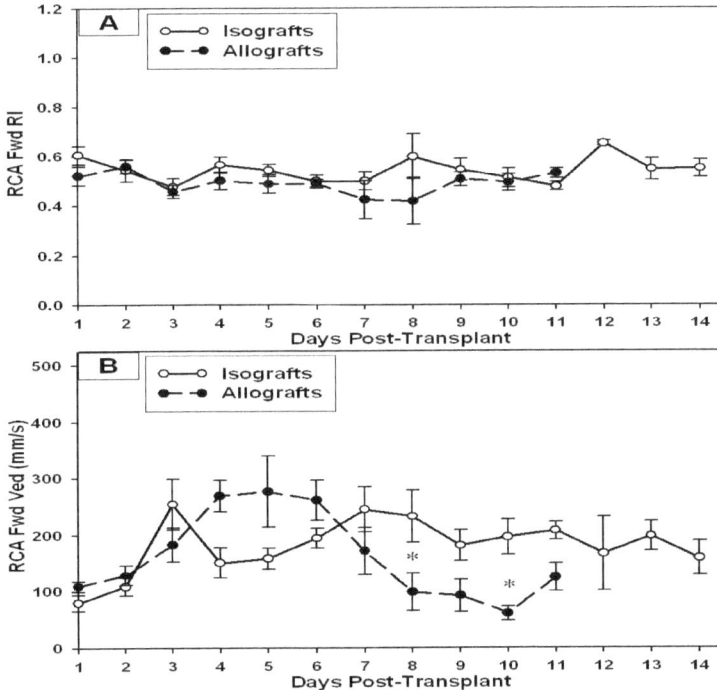

Figure A7 – Forward Resistance Index (RI) and End-Diastolic Velocity (Ved) in the Right Coronary Artery. A. No statistical differences in RI were detected. RI values for both groups (0.4-0.7) were similar. No data is available for days 12-14 for allografts because of the inability of determining Ved in from the RCA waveforms of rejecting grafts. B. Increases are seen early post-transplant in both groups with isografts peak at day 3, temporarily decrease and resume values close to ~200 mm/s. Allografts reach a maximal peak on days 4-7 and gradually decrease thereafter, significantly differing from isografts on days 8 and 10. Missing data for the latter days is due to the same reason given above. The data was analyzed using two-way repeated-measures ANOVA followed by post hoc testing by the Holm-Sidak method. Data are reported as means ± SEM. (*) indicates significance at p<0.05 between iso- and allografts.

No statistical differences in RCA RI were found and day 12-14 data for allografts was absent because of the difficulty of determining Ved in rejecting grafts. Allograft Ved decreases after day 6 and only transient differences were detected between the two groups of grafts.

158

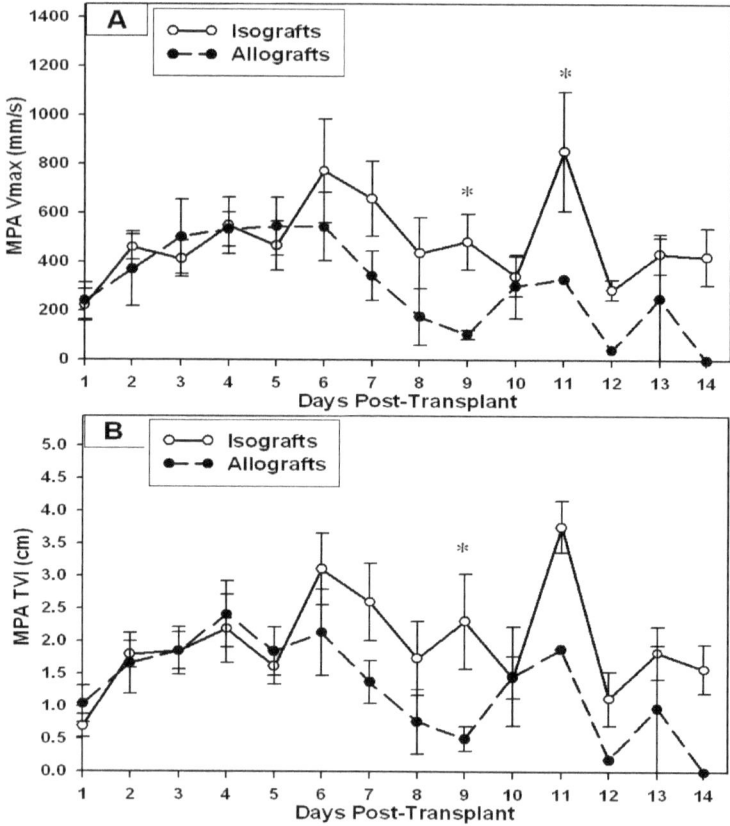

Figure A8 - Maximum Flow Velocity (Vmax) and Time-Velocity Integral (TVI) in the Main Pulmonary Artery. A. MPA Vmax. **B.** MPA TVI. Data are reported as means ± SEM. (*) indicates significance at $P = 0.041$ and $P=0.019$ for Vmax and TVI respectively between iso- and allografts.

MPA Vmax was similar for both iso- and allografts for days 1-5; isografts achieve generally higher values and allografts decrease variably after day 6. Transient differences detected on days 9 and 11 only (*$P=0.004$). TVI yields a similar temporal pattern for both groups of grafts in comparison to Vmax values.

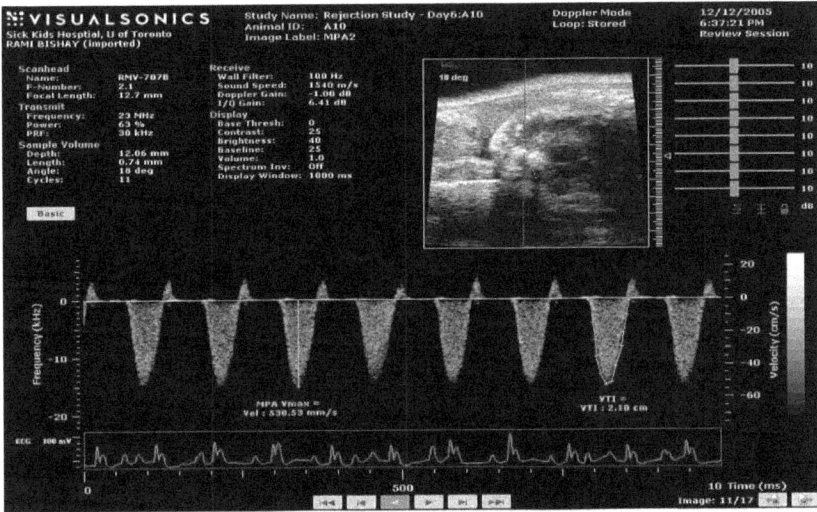

Figure A9 – Two-dimensional B-mode Acquisition and Doppler Quantitation of Maximal Main Pulmonary Arterial Flow. From the long-axis view (LAV) of the LV, a 90° rotation yielded the short-axis view (SAV) of the graft. Sweeping the transducer towards the base, visualization of the RV outflow tract and main pulmonary artery (MPA) was achieved. Doppler pulmonary arterial flow spectrum was recorded at the mid-point between the pulmonary orfice and anastomosis of the MPA with the recipient IVC. The 2-D image and Doppler spectra were taken from a Day 6 post-transplant allograft. *MPA*, Main pulmonary artery; *Vmax*, maximum velocity; *VTI*, Time-velocity integral.

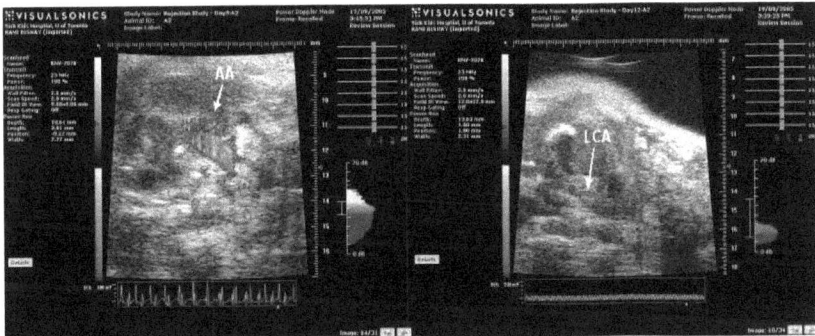

Figure A10 – Power Doppler of Aortic and Left Coronary Flow

160

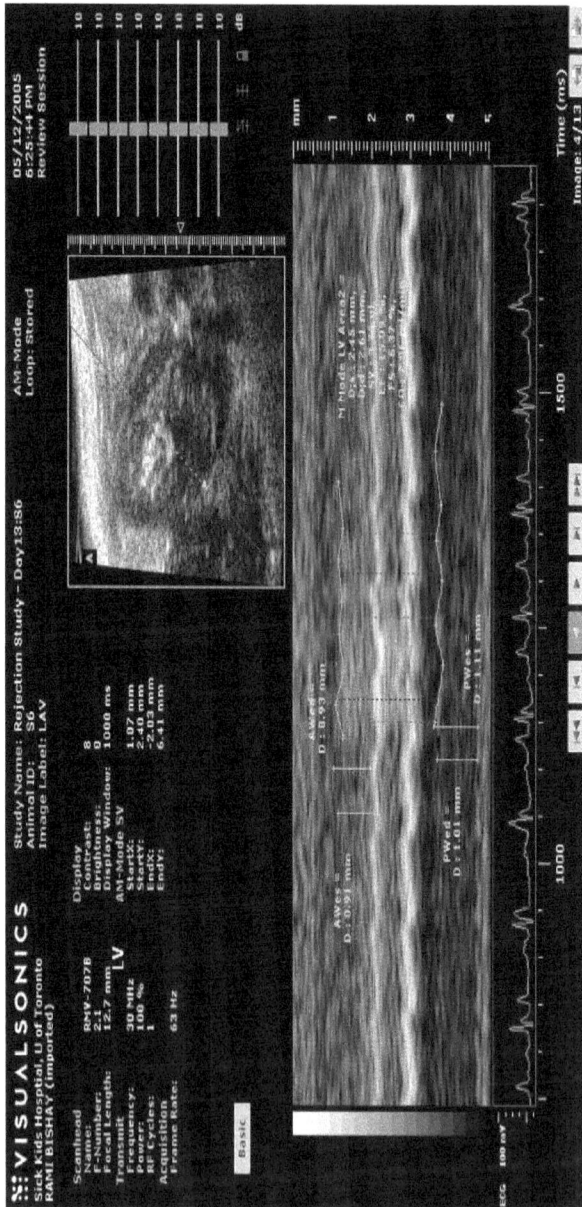

Figure A11 – Anatomical M-mode of the Left Ventricle: Anatomical M-mode of an isograft day 13 post-transplant. Due to limited free movement within the abdomen, a perpendicular cursor was not always possible when attempting to acquire M-mode derived parameters. In such cases, anatomical M-mode was utilized to analyze the grafts.

Anatomical M-mode allows for the acquisition of time course M-modes without the requirement for a perfectly perpendicular angle between the cursor and the tissue interface, which was advantageous for this animal model given the subtle abdominal movements of the cardiac grafts *in vivo*.

161

www.ingramcontent.com/pod-product-compliance
Lightning Source LLC
Chambersburg PA
CBHW070723220326

41598CB00024BA/3272